'*Being a Therapist in a Time of Climate Breakdown*' deftly uncovers the psychic ruptures and entanglements embodied by people's experiences of living with climate change. It highlights the demands that these experiences raise for contemporary therapeutic practice. A beautiful synergy of head and heart, this book combines arresting voices of lived experience with the wisdom of seasoned psychologists to provoke, challenge and enlighten. I recommend it for all who wish to acquaint themselves with psychology's potential to support healing, empowerment, and deep social transformations for sustainability.'

**Dr Charles Ogunbode**, *Associate Professor, University of Nottingham; Associate Editor at Global Environmental Psychology*

'Climate-aware therapists hold the most terrifying thoughts and feelings that people are experiencing on the planet today, as personal histories of trauma and hardship collide with escalating climate traumas and institutional abuses of power. This volume is a dazzling investigation of how these specialists care for themselves, their clients, and the world, as well as how therapists can help raise collective critical consciousness that aids people in not only coping but acting in turbulent times.'

**Britt Wray, PhD,** *Author of* Generation Dread *and Director of the Special Initiative on Climate Change and Mental Health in the Department of Psychiatry and Behavioral Sciences, Stanford Medicine*

'As we navigate the many challenges of the Anthropocene, *Being a Therapist in a Time of Climate Breakdown* couldn't be more timely. It is an essential and insightful resource for practitioners and citizens alike.'

**Tori Tsui** *is a climate justice organiser from Hong Kong who is now based in Bristol and the author of* It's Not Just You *and organiser with Unite for Climate Action, EarthPercent, Climate Live and Stop Rosebank*

'The book offers evidence and insights from climate activists and mental health experts emphasising how climate change impacts people's mental health. I appreciate how this book acknowledges the voices of those who disproportionately face the brunt of climate change and the injustices that it increasingly creates. Indeed, mental health professionals have a vital role in collectively addressing the existential threats of climate change, empowering people, and bringing hope to the future.'

**John Jamir Benzon R. Aruta, PhD,** *Environmental Psychologist in the Global South, Associate Professor, De La Salle University, Philippines*

'Engaging with global climate change is a new and wide territory for mental health professionals. For those seeking experienced guides, the authors in this timely volume provide an inspiring range of ethical, clinical, social, political and spiritual perspectives on climate-aware practice.'

**Thomas Doherty, PsyD,** *Licenced Psychologist, USA*

'*Being a Therapist in a Time of Climate Breakdown* is a groundbreaking, well-grounded and utterly necessary book. Practitioner's chapters intertwine with activist voices to produce a multi-faceted exploration of mind/body responses to climate and ecological breakdown, steeped in honesty, compassion, wisdom and radical insights. Out of this emerges a lively exploration of the way psychotherapies rooted within ecological consciousness and systemic analyses open up possibilities for mutually entwined healing for psyche, society and planet. A deeply moving and inspiring read.'

**Dr Sally Gillespie**, *Jungian Psychotherapist, member of Psychology for a Safe Climate, Australia, author of* Climate Crisis and Consciousness

'This is a thoughtfully curated, thought- provoking and highly accessible collection of works from many different voices on being alive as a human and therapist at this time. Facing into the reality and impacts of climate change, ecological devastation and social injustice is not easy but we do have a role to play – and that means rethinking a lot of what we took for granted in our practice and who or what it is we serve in the 21st century. A hugely enriching, enlivening book.'

**Linda Aspey**, *BACP Fellow, member of the Climate Psychology Alliance, founding member of the Climate Coaching Alliance, and co-editor of* Holding the Hope: Reviving Psychological & Spiritual Agency in the Face of Climate Change

'The current ecological crises and climate breakdown are symptoms of our inability to evolve our human consciousness. All Psychotherapists, psychologists and coaches need to turn their skills to this urgent task and this book has much to offer in this endeavour. Do read it and evolve your practice.'

**Professor Peter Hawkins**, *Co-author of* Ecological and Climate Conscious Coaching *and founder of the Bath Centre for Psychotherapy and Counselling and Renewal Associates*

'This is an essential read for trainee therapists, providing information and tools to respond to the climate emergency both inside and outside the therapy room. The book combines scientific facts, cutting-edge research, and moving personal accounts ranging from young children to practitioners, to meet the urgent need to become climate-aware therapists. The wide diversity of voices, from mapping eco-anxiety and climate grief, to offering approaches for hope and activism, will act as an indispensable guide as we navigate an uncertain future.'

**Laura Speers**, *Trainee, Bath Centre for Psychotherapy & Counselling*

'This brilliant work is not only for therapists but is also a worthwhile read for all literate humans in a time of climate breakdown. It is comprehensive and reaches into the depth of ecology and the human psyche. For our network "Global Climate Psychology for a Just Future" it is a real treasure to have.'

**Dr med Monika Krimmer**, *Psychoanalyst, Psy4Future, Germany*

'*Being a Therapist in a Time of Climate Breakdown* is a necessary response to an uncertain and precarious present and future. There is an ethical imperative that runs through this book, which callings upon individual therapists and their regulators to face up to the effects of climate and environmental crisis and to then act. In braiding together the insights of therapists and the voices of those beyond the therapeutic professions, a kaleidoscope of diverse knowledges emerges, which is underpinned by a commitment to naming the relationship between climate crisis and social injustice. This is then a timely book for all therapists that will assist their development and work with clients.'

**Dr Jamie Bird**, *Art Therapist and Senior Lecturer, University of Derby, UK; author of* Social Action Art Therapy in a Time of Crisis

'What a cornucopia of voices, perspectives, concepts and practices and yet *Being a Therapist in a Time of Climate Breakdown* at the same time manages to maintain a marvellous coherence! This wonderful volume demonstrates the breadth, vitality, intellectual rigour and critical edge of climate psychology. It will be a crucial resource in the difficult times to come, and not just for therapists and mental health professionals but also for engaged citizens seeking to make sense of their own experience of climate breakdown.'

**Paul Hoggett**, *Co-founder of CPA, Psychoanalytic Psychotherapist, Emeritus Professor of Social Policy, UWE*

'*Being a Therapist in a Time of Climate Breakdown* is invaluable for the expanding field of "climate aware" clinicians. Contributors to this multi-generational, multi-disciplinary collection address the political, ideological, and psychological contradictions embedded within the profession: the individualism, the pathologizing, the erasure of the more-than-human world, the untenable boundary between the "personal" and the "political". With conviction, the editors urgently call for and offer guidance in professional repurposing – helping us all "stay present" to the planetary crisis.'

**Rebecca Weston, LCSW & JD** *and Co-President, Climate Psychology Alliance – North America*

# Being a Therapist in a Time of Climate Breakdown

This book introduces readers to the known psychological aspects of climate change as a pressing global concern and explores how they are relevant to current and future clinical practice.

Arguing that it is vital for ecological concerns to enter the therapy room, this book calls for change from regulatory bodies, training institutes and individual practitioners. The book includes original thinking and research by practitioners from a range of perspectives, including psychodynamic, eco-systemic and integrative. It considers how our different modalities and ways of working need to be adapted to be applicable to the ecological crises. It includes Voices from people who are not practitioners about their experience including how they see the role of therapy. Chapters deal with topics from climate science, including the emotional and mental health impacts of climate breakdown, professional ethics and wider systemic understandings of current therapeutic approaches. Also discussed are the practice-based implications of becoming a climate-aware therapist, eco-psychosocial approaches and the inextricable links between the climate crises and racism, colonialism and social injustice.

*Being a Therapist in a Time of Climate Breakdown* will enable therapists and mental health professionals across a range of modalities to engage with their own thoughts and feelings about climate breakdown and consider how it both changes and reinforces aspects of their therapeutic work.

**Judith Anderson** is a Jungian psychotherapist and psychiatrist who has been focussed on the psychological aspects of climate change since 2005. Part of the founding group of Climate Psychology Alliance in the UK, she became Chair of its Board of Directors from 2020. She is passionate about the need for psychotherapists to become climate aware.

**Tree Staunton** is a UKCP Honorary Fellow, and an Emeritus member of Bath Centre for Psychotherapy and Counselling (BCPC). She has been a Registered Body Psychotherapist, Supervisor and Trainer for over 30 years. She was Director of BCPC from 2011–2023 and continues to teach and supervise trainee psychotherapists. She is a long-standing member of Climate Psychology Alliance, UK.

**Jenny O'Gorman** is a queer, disabled Psychodynamic Counsellor from a working-class background. Based in the UK, Jenny works in the charities and community-based counselling sector with adults and adolescents. They sit on the Board of the Climate Psychology Alliance and on the Organising Committee of Birkbeck Counselling Association's monthly psychodynamic forum. Alongside counselling and consulting work, Jenny is a writer and climate justice activist.

**Caroline Hickman** is a psychotherapist in clinical practice and lecturer at the University of Bath, UK, researching children and young people's thoughts and feelings about the climate and biodiversity crisis internationally. She is co-lead author on a 2021 global study of 10,000 children and young people's feelings on climate change published in *The Lancet Planetary Health*.

# Being a Therapist in a Time of Climate Breakdown

**Edited by Judith Anderson, Tree Staunton, Jenny O'Gorman and Caroline Hickman**

Routledge
Taylor & Francis Group
LONDON AND NEW YORK

Designed cover image: © Will Baxter

First published 2024
by Routledge
4 Park Square, Milton Park, Abingdon, Oxon OX14 4RN

and by Routledge
605 Third Avenue, New York, NY 10158

*Routledge is an imprint of the Taylor & Francis Group, an informa business*



*British Library Cataloguing-in-Publication Data*
A catalogue record for this book is available from the British Library

*Library of Congress Cataloging-in-Publication Data*
Names: Anderson, Judith, 1952- editor. | Staunton, Tree, 1955- editor. | O'Gorman, Jenny, 1988- editor. | Hickman, Caroline, 1961- editor.
Title: Being a therapist in a time of climate breakdown / edited by Judith Anderson, Tree Staunton, Jenny O'Gorman and Caroline Hickman.
Description: Abingdon, Oxon ; New York, NY : Routledge, 2024. | Includes bibliographical references and index. |
Identifiers: LCCN 2023047675 (print) | LCCN 2023047676 (ebook) | ISBN 9781032565590 (hardback) | ISBN 9781032565606 (paperback) | ISBN 9781003436096 (ebook)
Subjects: LCSH: Climatic changes--Psychological aspects. | Therapist and patient.
Classification: LCC BF353.5.C55 B456 2024 (print) | LCC BF353.5.C55 (ebook) | DDC 155.9/15--dc23/eng/20240207
LC record available at https://lccn.loc.gov/2023047675
LC ebook record available at https://lccn.loc.gov/2023047676

ISBN: 978-1-032-56559-0 (hbk)
ISBN: 978-1-032-56560-6 (pbk)
ISBN: 978-1-003-43609-6 (ebk)

DOI: 10.4324/9781003436096

The Open Access version of this book was funded by Climate Psychology Alliance.

Dedicated to all our friends and colleagues in the Climate Psychology Alliance, to the young people growing up within the ecological emergency and to all those working towards a just future.

# Contents

# Acknowledgements

With thanks to our colleagues in CPA and throughout the psychological professions, to our clients, to our friends, families and communities, for all the care, thinking and feeling which has supported us in writing this book.

# Contributors

**Garret Barnwell** is a South African clinical psychologist and community psychology practitioner. In addition to providing psychotherapy, he has worked on health justice issues in detention and extractive settings for Doctors Without Borders; the youth-led #CancelCoal case for the Centre for Environmental Rights; and the psychological impacts of opencast coal mining for All Rise Attorneys for Climate and Environmental Justice. He is also a research associate at the University of Johannesburg's Psychology Department.

**Steffi Bednarek** is a UK Psychotherapist, editor and consultant on climate psychology, systems thinking, and trauma-informed change. Her work explores the intersection between climate change and mental health. She is a 'firekeeper' at the worldethicforum, member of the American Psychological Association's Climate Change Community of Practitioners and member of the Climate Psychology Alliance. She has managed mental health services and has consulted national governments, the corporate sector, global financial institutions and large NGOs.

**Gillian Broad** has been involved with Climate Psychology Alliance (CPA) for the past three years. Since joining she has worked closely with Rebecca Nestor, founder of CPA climate cafes, to develop the climate café work and its reach across the UK. The contexts range from voluntary sector mental health organisations to higher education institutions hosting climate cafes with diverse disciplinary groups. She is constantly in awe of the powerful impact on the wellbeing of individuals of sharing eco-related feelings in group settings.

**Paula Conway** is a psychoanalytic psychotherapist in the UK with over 30 years' experience working with children and adolescents before training as an adult psychotherapist at the Tavistock Clinic. She ran mental health services for children in care, post-adoption support in Kent, and set up Grow2Grow, a farm-based ecotherapy service. She has taught extensively and published on applied psychoanalysis. She now works in private practice in East Sussex, and with her husband, set up Hides Farm Therapy Centre.

**Leslie Davenport** is a climate psychology consultant and therapist in the USA exploring the intersectionality of climate, education, policy and social justice. She has authored four books, including *All the Feelings Under the Sun* (American Psychological Association, 2021). She is an advisor to the Post Carbon Institute, Project InsideOut, Climate Mental Health Network, One Resilient Earth and on faculty and co-lead of the Climate Psychology Certification at the California Institute of Integral Studies. www.lesliedavenport.com.

**Iona Fredenburgh** is a co-director, senior faculty member and trainer at Processwork UK. She has a background in political and social justice issues, and in subtle energy awareness. For many years she has facilitated groups that support people to reconnect deeply with nature and spirit.

**Kelvin Hall** practised Integrative Psychotherapy in the UK for 36 years, latterly including outdoor and equine-assisted work in his approach. He has regularly contributed articles on the significance of encounters with other-than-human life to various publications, and given talks and workshops on a number of training programmes. He has received an Emeritus award from the Bath Centre for Psychotherapy and Counselling (BCPC). He is also a keen horseman and professional storyteller.

**Jo Hamilton** is currently a lecturer and a Postdoctoral research impact fellow in the Geography department at the University of Exeter. She researches the equity and psychosocial dimensions of public engagement with climate change and energy transitions in the UK. Alongside research, her background includes climate change engagement, supporting community climate change and energy projects, facilitating group work processes such as the Work that Reconnects and being a musician.

**Wendy Hollway** is Emeritus Professor of Psychology at the Open University, UK. Recent writing includes an OpenLearn (OU) free online course, *Climate Psychology*, with Trudi Macagnino; a book, *Climate Psychology: A Matter of Life and Death* with P. Hoggett, C. Robertson and S. Weintrobe (Phoenix Publishing House Ltd, 2022); and a chapter 'Putting the "eco" in psycho-social' (in S. Frosh et al. (eds) Palgrave Handbook of Psychosocial Studies, 2022).

**Chris Johnstone** is co-author, with Joanna Macy, of *Active Hope*, a widely-known book about the Work That Reconnects, and author of *Seven Ways to Build Resilience* (Robinson, 2019). He has a background in medicine, psychological therapies, groupwork and coaching, and worked for many years in the UK as a group therapist in a specialist mental health team. His online trainings for resilience and active hope reach thousands of people through websites at https://collegeofwellbeing.com and https://activehope.training.

**Rosie Jones** is a facilitator and trainer in The Work That Reconnects, based in the UK. She has a background as a clinical psychologist, having worked in NHS mental health settings and private practice. She now focuses on initiating or supporting grassroots projects, particularly where land and community regeneration are combined. At Psychology for Ecology she enjoys sharing processes and skills that can support the healing and flourishing of human and non-human life.

**Peter Kalmus** is a climate scientist at NASA's Jet Propulsion Laboratory in the USA, speaking on his own behalf. He is the recipient of NASA's Early Career Achievement Medal and Exceptional Scientific Achievement Medal. His research includes projections of extreme humid heat and projections of ecosystem breakdown. Peter is an activist, organiser and supporter of numerous climate direct action groups around the world. He has been arrested for nonviolent climate disobedience.

**Trudi Macagnino** is a psychotherapist in private practice in the UK. She has an interest in our relationship with the more-than-human world. She is an academic staff tutor in Psychology & Counselling with the Open University where she is involved in the delivery of counselling modules. She is a member of the Climate Psychology Alliance and is co-authoring a free Climate Psychology course for the Open University. She is currently writing up her PhD research.

**Julian Manley** is Professor of Social Innovation at the University of Central Lancashire, UK and a Director of both the Centre for Social Dreaming (CSD) and the Climate Psychology Alliance (CPA). Publications include the 2018 book *Social Dreaming: Associative Thinking and Intensities of Affect* (Springer, 2018) and the article 'The jewel in the corona: Crisis, the creativity of social dreaming, and climate change' (in *The Journal of Social Work Practice*, Vol. 34/4, 2020).

**Jo McAndrews** works in the UK with young people, parents and professionals who are engaged with children, particularly around growing resilience in the face of crisis. She is particularly interested in the impact of systemic trauma on child development, and the role of care culture in growing resilience in children, adults and communities. She is a member of the CPA, developing support for young people and the adults around them. More at www.jomcandrews.com

**Sue Milner** is a senior faculty member of Processwork UK. She is a co-founding member of the Equality, Diversity and Intersectionality working group within the Humanistic and Integrative Psychotherapy College of UK Council for Psychotherapy (HIPC), and has an MA in social work. Her deep love of nature has led her to also lead vision quests.

**Gareth Morgan** is Academic Director of the University of Leicester's Clinical Psychology Doctorate. He is co-chair of the Association of Clinical Psychologists' Climate Action Network, based in the UK and member of the Climate Psychology Alliance and XR Psychologists.

**Rebecca Nestor** is a facilitator and organisational consultant in the UK whose main focus is supporting those facing the climate crisis. She is a member of the board of the Climate Psychology Alliance, adjunct faculty at the Boston (US) Graduate School of Psychoanalysis and a visiting lecturer at the Tavistock and Portman NHS Foundation Trust, where she contributes to the masters and doctoral programmes on organisational consulting and leadership.

**Panu Pihkala**, from the University of Helsinki, Finland, is an expert in interdisciplinary eco-anxiety research. In addition to writing books and research articles, he works as a workshop facilitator. Among other positions of trust, Pihkala serves as an advisor for the Finnish national project on social and health sector responses to eco-anxiety (www.ymparistoahdistus.fi). He hosts the podcast Climate Change and Happiness, together with Dr Thomas Doherty, and often co-operates with artists and educators.

**Halina Pytlasinska** is a UK-based psychotherapist and relationship therapist, clinical supervisor and consultant trainer. She facilitates dream sharing in groups to enhance ecological awareness and decolonising work. Through reading Black Feminism and listening to indigenous voices, Halina is exploring how a kinship worldview enhances recovery from social and racial conditioning. Her group work integrates bodywork, creativity, storytelling and guided visualisation to engage the imagination and connect with earth wisdom.

**Chris Robertson** has been a psychotherapist in the UK since 1978. He studied child psychotherapy, psychosynthesis, archetypal psychology and family therapy. He co-founded ReVision, an integrative psychotherapy training with soulful perspective in 1988 from which he is now retired. From 2018–2020 he was chair of the Climate Psychology Alliance with whom he still works offering workshops. Publications include: co-author of *Climate Psychology: A Matter of Life and Death* (Phoenix Publishing House Ltd, 2022) and *Culture Crisis* in Depth Psychology and Climate Change (Routledge, 2020).

**Nick Totton** is a body psychotherapist, trainer and supervisor, currently living in Sheffield, UK. He has published several books, including *Psychotherapy and Politics* (Sage, 2000); *Embodied Relating* (Routledge, 2018); *Wild Therapy* (PCCS Books, 2011); *Different Bodies* (PCCS Books, 2023), and *Sailing to Bohemia* (Magpie Moon, 2023). He was founder-editor of

*Psychotherapy and Politics International* (John Wiley and Sons Ltd), and has been involved in starting and/or running several therapy organisations, including Psychotherapists and Counsellors for Social Responsibility, the Psychotherapy and Counselling Union and the Body Psychotherapy Network.

**Celia Turley** is an artist-facilitator and cultural organiser experimenting with answers to the knotty question of how collaborative, creative work can imagine and practice possibilities for collective care, transformation and repair and planetary survival. She steers The Resilience Reading Circle, studies plant medicine and organises public programmes and social art projects in Bristol, UK and beyond.

# Foreword

What an honour to be asked to introduce this important book from my dear British Climate Psychology Alliance colleagues and friends. My own work, *Climate Crisis, Psychoanalysis, and Radical Ethics* (Routledge, 2015), written nearly ten years ago, seems to me fearfully out of date now, but surely shares the spirit you will find in this present book. These years of fearful and fearless thinking, of more-than-courageous activism, of teaching, but most of all of acceleration of the climate crisis into an emergency, have made it clear that the urgency many of us have long felt was not misplaced. Now you can read one passionate expert after another: on the emergency we are inhabiting, on the meanings – both practical and ethical – we are coming to realise, on the imperatives for everyone, including especially psychotherapists of all persuasions.

Clearly these essays come from long and serious conversations fostered within and around CPA. They will surely generate more questions and debates around the nature of our crisis, around its several places in our clinical work and around the ways we psychotherapists can and must function in our lives as citizens. We can no longer isolate ourselves in our professional concerns and disputes while Rome burns. Every reader will be made uncomfortable, at the very least, facing demands that many of us did not consider as we chose or grew into our professions. This discomfort may grow into huge sorrow and/or outright rage, not to mention guilt, but all of these can show ethical awakening at the realisation of what our generation is leaving to the next ones. Even worse perhaps, we see what indigenous people, like those of Lahaina, Maui, have suffered from neglect while wealthy tourists used up their water. Once again, we are forced to remember that while we are all suffering the effects of our heating-up planet, those already disadvantaged by colonialism and slavery are suffering much more.

I am grateful to the editors and authors of this new book for all the work you are doing to keep our ethical eyes open.

**Donna M. Orange, PhD,** *New York University Postdoctoral Program in Psychoanalysis and Psychotherapy (faculty and consultant)*

# Introduction

*Judith Anderson, Tree Staunton, Jenny O'Gorman and Caroline Hickman*

> *It is an intertwining of Science, Spirit and Story – old stories and new ones – that can be medicine for our broken relationship with the earth.*
>
> Robin Wall Kimmerer, Braiding Sweetgrass (2013: x)

What meaning can be found within the current overlapping and unfolding crises – social, cultural and environmental – and how can the psychological professions offer insight, increase awareness and contribute to the vital consciousness raising project which is currently developing?

This is a book for therapists, and a book for anyone who seeks to understand the psychological roots of our collective apathy and continued inaction in relation to life-threatening climate breakdown. We have been warned that it is vital that we *act now* to avoid catastrophe, and yet even knowing this we do not seem able to find the necessary intentions or resources. Paul Hoggett (2019) questions if it is possible for therapists to remain neutral in the face of climate catastrophe, given that we ourselves are part of this increasingly heated, disrupted and frightening world. Is neutrality desirable or even possible in a world in crisis?

Why this book, and why now? In *Whose Story Is This?* Rebecca Solnit writes 'This is a time in which the power of words to introduce and justify and explain ideas matters, and that power is tangible in the changes at work. Forgetting is a problem … words matter, partly as a means to help us remember' (2019: 3). For us it represents another mark in the sand, an attempt to gather up important pieces of the story, and lay them before ourselves and others – scientists, psychologists, therapists, health workers and storytellers – as we attempt to make meaning. Our hope is for it to spread its growing message like mycelium, touching the roots of our collective psyche and undermining the concrete structures we have built around our egos. Our message to the therapeutic community is that *every therapist must become a climate aware therapist.* We hope it will become clear through these chapters what that involves.

DOI: 10.4324/9781003436096-1

The language we use matters, and we are conscious that to those less familiar with the detail of what science and lived experience show us about the consequences of climate breakdown, talk of such all-encompassing loss may feel like hyperbole. Sadly, it is not. Throughout this book we will refer to climate breakdown, ecological destruction and biodiversity loss. We see all the varied aspects of planetary boundary breaches (Rockström et al., 2009) as parts of what is overall colloquially called climate change.

In 2022 the IPCC Report first addressed the mental health impact of climate change. Prior to this, the emphasis was on physical health impacts. The recognition of post-traumatic stress disorder, trauma, depression and anxiety validated what many have understood and felt for some time: that the effect of the climate and biodiversity crises on mental health is significant, and widespread. At the report's launch, António Guterres, the Secretary-General of the United Nations responded with 'Today's IPCC report is an atlas of human suffering and a damning indictment of failed climate leadership... delay means death … I know people everywhere are anxious and angry; I am, too … every voice can make a difference … now is the time to turn rage into action' (2022).

A number of our authors refer to the concept of resilience. We acknowledge that 'resilience' has been used in a colonial context, individualising societal issues in the same way that fossil fuel companies created the concept of carbon footprints to divert from the need for systemic changes. It can imply a defensive position where one remains resolute and untouched by societal harms. However, within the context of climate psychology, resilience refers to being able to integrate multiple different states, including the hypocrisies inherent in living within a 'culture of uncare' (Weintrobe, 2021). Resilience in this sense is more akin to that which emerges through biodiversity – a way of being wherein diverse emotions are held simultaneously and shocks can subsequently be more effectively absorbed. It is our hope that the chapters in this book can be integrated in such a way as to provide a kind of therapeutic resilience.

Within this, our position is that the breakdown of the climate does not just refer to the established weather patterns but also to the social climate, which emerges in politics, culture and relationships. Ecosystemic thinking proposes a complex of relationships between all living beings and their environments as a dynamic and intersectional system. Humans are one part of a larger, connected web; changes to any aspect of this system ripple through to impact everything else within the system, and no part exists in isolation. When we write of ecological destruction, it is this interwoven tapestry of life which forms the frame, being object, subject and location of large-scale abrupt change. The multiplicity of lifeforms and varied habitats are taken together in our use of the term biodiversity. Chapters in this book speak to the felt impact of the loss of species-rich surroundings, including a reduction in diverse ways of thinking about and responding to this devastation.

The editors are all members of Climate Psychology Alliance (CPA[1]), an organisation which launched in 2011 and has become a membership organisation with an international reach. Its purpose is to explore psychological responses to the climate crisis in order to strengthen relationships and resilience for a just future. It researches and publishes, develops methods of therapeutic support and provides CPD for psychological professionals. As a group, we see the climate and ecological crisis as inextricably linked with racism, coloniality and social injustice.

The authors and contributors in this book are drawn from a diverse spectrum of 'ways of being human' in our current times. Whilst acknowledging our limitations of representation, we have considered ethnicity, disability, class, gender, sexuality, neurodiversity and age in bringing together an interdisciplinary range of authors and voices to explore what it means to be a therapist, or to engage with therapy during a time of climate and ecological breakdown. We choose to include discussion of the academic, therapeutic, political, emotional, cognitive and imaginal to hear the voices of both therapists and of people who are not therapists but have something important to say about how therapy and the climate crisis intersect with and affect their lives. Through including their voices, we aim to address the importance of listening to and respecting other voices alongside those of therapists and mental health professionals.

This book calls the profession to account, addressing the regulatory bodies and training institutes as much as the individual practitioners. In March 2014, as members of the UK Council for Psychotherapy (UKCP[2]) *Diversity, Equality and Social Responsibility Committee* (DESRC), two of the editors of this volume, Judith Anderson and Tree Staunton, put forward a proposal for an *Environmental, Sustainability and Climate Change Policy*. It suggested that the UKCP show leadership in its own sphere of influence in recognising the important role that the psychological professions have to play, and may need to play in the future, by promoting and supporting conversation and debate among members. Further suggestions included that it develops links with relevant organisations and committed individuals to share ideas and liaise with other organisations to educate and campaign for governmental policy change.

Other regulatory bodies within the psychological professions followed; BPS[3] created a climate change working party, and BACP[4] developed policies around energy consumption and waste management. However, professional development offered by BACP in terms of publications, training and events has been led by focused initiatives from within the membership rather than by policy from the centre.

More recently in 2021, UKCP initiated a cross-sector collaboration, the Climate Minds Coalition, with strong links to the *Climate Cares*[5] project, yet it has required member led initiatives to bring attention to the need for professional development and changes to training standards.

Throughout this book clinicians, academics and teachers of the profession emphasise the need to update our practice with climate in mind.

The position of the regulatory bodies with regard to arrests and public order convictions related to climate action is currently being tested, and so far, Fitness to Practise cases brought against practitioners have not resulted in removal from registers. This is becoming a very real dilemma when practitioners feel conscience-bound to act. Chapter 3 considers professional ethics in light of our responsibility as citizens.

It has been the experience of psychotherapy trainees that discussing social justice issues within therapy sessions has been labelled as 'grandstanding' and 'bringing politics into the therapy room'. Yet regulatory bodies increasingly set requirements for practitioners to update their awareness in practice with regard to issues of race, culture, sexuality and gender, implying an engagement with the social and systemic dynamics that perpetuate prejudice and inequalities. Intersectionality of all these combines to promote a higher level of awareness and offers the opportunity for consciousness raising. In proposing this book, some early feedback included the sentiment that it was difficult to see the relevance of climate breakdown within the therapy room. Throughout the chapters, our authors repeatedly demonstrate that not only is it relevant, but it is vital for us to make space for ecological disturbance to enter our practice.

It is noted throughout this book that when clients turn to therapy for help in dealing with one of the biggest existential crises humanity has faced, they may be failed, missed, let down or diverted. It can seem that the therapist's response to this potentially overwhelming and engulfing reality has been inadequate, and occasionally even damaging. However, it would be wrong to suggest that this is an accurate representation of the overall response of professionals – we hear more about these failures when clients turn to the therapeutic support services of CPA. Currently, many therapists are struggling to articulate and mediate their own position in relation to their clients presenting these issues, feeling the emotional impact of listening to the environmental harms being caused by frequent flying and extravagant lifestyles at one end of the spectrum, whilst also supporting those clients whose environmental actions lead them to arrest, suffering and deprivation.

This is particularly the case with children and young people who, whilst having some agency, in effect have little power to influence change at a political level in the world and are largely dependent on adults to understand and take action on their behalf. UNICEF in 2021 identified that the climate crisis is a child rights crisis in the children's climate risk index. They stated that almost every child on earth (over 99%) is exposed to at least one of the major overlapping climate and environmental hazards, shocks and stresses. The long-term mental health impact on children can be understood through stress-vulnerability models of health.

Because of chronic stress in childhood, there is a long-lasting impact that increases the risk of developing mental health problems in adolescence and adulthood. The effects of climate change on children have to be seen as one of the greatest injustices in the world today. It often happens that children are seen through the eyes of the adults around them, rather than directly through their own understanding. It has been argued that children's voices should be at the heart of discussions about climate change, as it is their future that we are shaping by our actions today. Children and young people's (CYP) voices repeatedly show up throughout this book, at its beginning, middle and end.

## The Voices

The Voices you will find in this book are from young people and adults, activists, mothers and artists, who recognise that they struggle with climate anxiety and have looked for help from various sources. They do not regard their anxiety as evidence of pathology but as a demonstration of their capacity to care. They were asked to respond to these questions: 'How does the climate crisis make you feel?', 'What do you want therapists to know?', and 'What do you want them to understand?' The Voices are unfiltered. Some may be uncomfortable, hard to read, see or comprehend, but they are authentic and from the heart. It can be difficult to be coherent or articulate, or know how to communicate your horror and despair, whilst also terrified that no one will understand. These contributors show great courage in finding ways to try to communicate this to us.

In recent years some young activists have taken extreme action to try to show how they feel, without words – setting themselves on fire, hunger strikes, suicide – not knowing how to live in a world that does not seem to care. We (and they) hope that their voices can be heard and make some difference here. All of them have faced some disappointment from therapy, as well as, or alongside comfort and validation.

The Voices pieces include contributions from Emily Kelsall – otherwise known as TRex vs TMX – who brilliantly illustrates her internal relationship between herself as a child, teenager and young adult in conversation about the climate crisis. Fehinti Balogun uses art to express his despair; he is an actor, artist, activist and powerful advocate for black people's experiences of climate breakdown. Eva Bishop and Chloe Naldrett speak as mothers and climate activists about what it means to them to parent with imagination and radical love, whilst facing their anxiety about keeping their children safe. Timothy Morton shares his view on climate trauma and the need for 'punk' activism; he extends into the depths of his own personal anxiety to show the reader how far he believes we need to all try to reach. Shelot Masithi speaks powerfully about living with regular water shortages in South Africa. She is founder of She4Earth and her contribution in this book is based on a keynote lecture for a CPA international summit. Helen Leonard-Williams and Elouise

Mayall, both activists with UKYCC,[6] describe their experiences as young people who have developed a mature and clear perspective on systemic injustice through being engaged in climate activism for some years. Having faced disappointment and navigated disillusionment, they have not lost their capacity to care.

We also hear from younger children in this section who have chosen to use abbreviated names or pseudonyms. T. M. Walshe talks about how it feels to strike from school for the climate every Friday, despite teachers and friends struggling to understand what it means to her. Frankie tells us how she feels about making a choice to follow a career in aviation at the age of 18, as her way of dealing with climate anxiety.

Will Baxter's colour photograph provides the cover image for this book. His love of the natural world represents everything we need to fight for and also risk losing. His narrative with the black and white image inside the book explains how many days it took him to get this photograph, and encapsulates his determination, vision, imagination and heart. We are delighted and humbled to be able to recognise this in him, as with all the contributors.

Finally, Maddie Budd shows her brutally honest feelings in her writing, which can be felt as fragmented, disjointed and even despairing, and all the more wonderful for this. We support her incoherence as an understandable narrative of eco-anxiety.

## The Chapters

In the first section, **The Trouble We're In**, Peter Kalmus, a US climate scientist with NASA, opens with his experience of communicating and acting on climate breakdown. In April 2022 he was arrested for protesting outside JP Morgan Chase Bank, leading funders of fossil fuel projects. Through his tears, he spoke about how scientists are not being listened to, saying: 'I am willing to take a risk for this gorgeous planet, for my sons … for all the kids of the world, all the young people, all the future people … this is so much bigger than any of us. We're going to lose everything.' He presents here some 'fact clusters' alongside his vital commentary.

Judith Anderson and Rebecca Nestor follow with an outline of the mental health and emotional impacts of climate breakdown and reflections on the profound and complex psychologies associated with inaction, avoidance and a clinging to 'business as usual' in high- and middle-income countries. Jenny O'Gorman focuses on our ethical codes and the ways in which the ecological emergency challenges our existing ethical frameworks. They consider how as practitioners we can respond to our clients, and how we can attend to our own changing personal and professional needs in these demanding times. They argue for a revisioning of therapeutic ethics in order to meet the challenges of these times. We encourage this debate.

We turn next to wider **Systemic Understandings** and offer three very different approaches. Steffi Bednarek questions the individualism inherent in our current therapeutic models, and points to their collusion with the capitalist system. She offers a reframing of psychotherapy in relation to the bigger picture, suggesting that the crisis may be a threshold moment for the evolution of the profession. Gareth Morgan gives us some important insights from the working group within the British Psychological Society on an important reconfiguring of mental health issues using the Power Threat Meaning Framework showing its relevance to supporting one another as we navigate the unprecedented challenges of climate and ecological breakdown.

Iona Fredenburgh and Sue Milner introduce the concept of 'deep democracy' and illustrate how if we are 'process led' in our work with groups and individuals, we can trace how the system lives within us, bringing an experience of the 'sentient unified field'. To end the systemic section, Celia Turley and Jo McAndrews report on *Motherhood in a Time of Climate Crisis*, a collaborative theatre-making process that gives voice to frank, personal stories of how the climate crisis shapes how, and if we mother. This unique project provides a space for the deepest feelings about motherhood, and bears witness to a necessarily radical care approach in all we do in this situation. Many of our contributors cite Sally Weintrobe's critique of the 'culture of uncare'. Turley and McAndrews show us its opposite.

The third section moves into the detail of **Becoming a Climate Aware Therapist**. Caroline Hickman's chapter on climate aware therapy with children and young people brings a wealth of insights from her practice and her internationally renowned research. She proposes a climate and biodiversity crisis lens through which therapy can respond to these concerns and proposes a range of approaches, drawing strongly on the imaginal. Trudi Macagnino draws on her PhD research interviews with both clients and therapists. She elaborates and discusses in detail what is meant by revisioning therapy as a collective eco-psycho-social endeavour rather than a purely individualistic one, focusing on the implications for practice.

Two case studies follow, as Paula Conway wonders with us if, in spite of overt climate silence, the process of mourning damaged inner 'objects' can facilitate the task of facing the greatest threat to humanity we have known. Garret Barnwell brings a synthesis of Lacanian and climate psychology perspectives to his work with a client in a manic psychotic state as a response to the climate crisis.

Panu Pihkala's chapter also belongs here. We welcome his extensive research experience focused on climate grief, an issue that is a repeated thread in the chapters. We find his emphasis and insights on the non-finite aspects of climate grief particularly relevant for practitioners. This section ends with Tree Staunton's chapter drawing together the regulatory work that has achieved compulsory training in ecological awareness, with the views of trainees and graduates about their need for additional training. A strong argument is made for the overarching need for embodied approaches in training and practice.

We then shift to an ecopsychological perspective, acknowledging an older, parallel discipline to climate psychology. In the section **The Ecological Self**, Kelvin Hall invites the reader to enter the 'Zone' in an encounter with other-than-human creatures, which he says has a particular effect on human consciousness, leading to a sense of being 'at home' on the earth. He offers examples from his own practice which illustrate the transformative effect this can have. Nick Totton reminds us that it is therapy itself which needs to change to embrace the ecosystem, and to 'take the wild back into the therapy room'. This means, he says, to invite ourselves and our clients to enter an ecological consciousness, a sense of connectedness that can arise when we relax and cease to impose control over what should happen. Leslie Davenport's chapter brings perspectives honed over many years in the field of health, bringing intuitive, somatic, creative and beyond-the-logical approaches to the transformation required facing the climate crisis. Finally, Chris Johnstone and Rosie Jones's chapter takes us through the spiral of *The Work That Reconnects,* a therapeutic practice, developed over 40 years out of the work with Joanna Macy. It aims to help people find a sense of agency in relation to their concerns about the world and suggests that our collective resilience may be strengthened through these practices.

In the final section, **Community and Social Approaches**, we explore methodologies that have been developed outside the consulting room by therapists and non-therapists. Julian Manley, Wendy Hollway and Halina Pytlasinska show how climate change manifests in the image-rich and shared tapestry of the language of social dreaming, whilst Gillian Broad presents an evocative description of 'Ways of Being' in a climate café. The reader who aspires to work with a climate organisation would be well-advised to equip themselves with the important insights from Rebecca Nestor on how the climate and ecological emergency impacts group dynamics.

Jo Hamilton delves deeply into the subject of grief, offering insights from her research into socio-ecological group practices which help us to think and feel into our connection with ourselves, others and the more-than-human world. Chris Robertson ends our book by addressing the psychological work involved in being with climate breakdown and the many and varied emotional states this evokes, challenging us to stay with these difficult truths and embody a more emergent practice.

None of the contributors to this book view climate breakdown as marginal – it is central to our concerns, and for some, central to our lives. We make no apology for this. All our writers in different ways acknowledge the systemic socio-political attitudes and injustices that have led to the state we are in. For that too we make no apology. Our distress is a measure of our capacity to care, and we feel both, equally.

Without dismissing our collective experience and that of the people we are working with, we do not claim to have the final word. Staying present to the climate and biodiversity crises requires us to continue to face our own vulnerabilities and struggle with the changing ecological landscape.[7]

We hope you will be moved by what you read and inspired to take up a position in your own work which recognises that a culture of care comes about through the interaction between personal, communal and ecological dynamics. We respect and are aware of those who have come before us, and hope many will come after. Just as 'every fraction of a degree matters' in the movement towards a liveable future (CARE, 2021), so too does every fraction of conversation and connection matter when it comes to the re-telling of our story and the re-making of our future.

## Notes

1 See www.climatepsychologyalliance.org for resources such as podcasts, journal, workshops, Therapeutic Support directory, and more.
2 The United Kingdom Council for Psychotherapy holds the UK's national register of psychotherapists and psychotherapeutic counsellors, and is the largest professional regulatory body in the UK.
3 British Psychological Society.
4 British Association for Counselling and Psychotherapy.
5 Climate Cares is a collaboration between the Institute for Global Health Innovation (IGHI) and The Grantham Institute – Climate Change and Environment which aims to equip individuals, communities and healthcare systems with the knowledge, tools and resources to become resilient to the mental health impacts of climate change.
6 United Kingdom Youth Climate Council.
7 This introduction was written in a week during which heat records were being broken almost daily, with 46°C recorded in Sardinia; Italian cities issued red alerts; wildfires spread across Europe and Africa; China recorded 52.2°C in Xinjian; in the US a heat dome over the south-west left tens of millions of people under extreme heat warnings, with New York obscured by smoke from Canadian wildfires; flooding affected thousands in Portugal, Spain, Germany, Italy and India; and UN Secretary-General Guterres pronounced that we have moved from global warming to global boiling.

## References

CARE Climate Justice Centre (2021) Response to IPCC: 'Every fraction of a degree matters to those already on the frontlines'. Available at: https://careclimatechange.org/care-on-ipcc-every-fraction-of-a-degree-matters-to-those-already-on-the-frontlines/ [Accessed 27 July 2023].

Guterres, A. (2022) Transcript of Remarks to the press conference launch of IPCC report. Available at: https://media.un.org/en/asset/k1x/k1xcijxjhp [Accessed 27 July 2023].

Hoggett, P. (2019) *Climate Psychology: On Indifference to Disaster*. London: Palgrave Macmillan.

Kimmerer, R. W. (2013) *Braiding Sweetgrass: Indigenous Wisdom, Scientific Knowledge, and the Teachings of Plants*. Minneapolis: Milkweed Editions.

Rockström, J., Steffen, W., Noone, K. et al. (2009) Planetary boundaries: Exploring the safe operating space for humanity. *Ecology and Society*, 14(2): 32.

Solnit, R. (2019) *Whose Story is This? Old Conflicts, New Chapters*. Chicago: Haymarket Books. Introduction: Cathedrals and Alarm Clocks.

UNICEF (2021) *The State of the World's Children 2021 – On My Mind: Promoting, protecting and caring for children's mental health*. Available at: www.unicef.org/reports/state-worlds-children-2021 [Accessed 27 July 2023].

Weintrobe, S. (2021) *Psychological Roots of the Climate Crisis: Neoliberal Exceptionalism and the Culture of Uncare*. New York: Bloomsbury.

Voice 1

# T-Rex vs TMX cartoon

*Emily Kelsall*

DOI: 10.4324/9781003436096-2

*Plate 1.1* Emily in conversation with herself as a child, teenager and young adult.

*Plate 1.2* Emily in conversation with herself as a child, teenager and young adult.

Emily Kelsall, Canada.

Section One

# The Trouble We're In

Chapter 1

# Facing Difficult Climate Truths

*Peter Kalmus*

This chapter gives a resumé of the science behind climate breakdown and its irreversible impact on all global life, particularly the global poor. It discusses the causes of climate change, warming trends, and the role of the fossil fuel and agricultural industries. Written by a climate scientist and activist, it calls for commensurate and immediate decisive action. Further, it suggests it is critical for those working in climate fields to seek psychological support *and* for therapists to be prepared to deal with intense climate issues.

I am a climate scientist. I think about the climate and ecological emergency every day – my mind cannot look away. I've been looking squarely at this emergency and grappling with it since 2006, the year I became a father. Having a child kicked me into a higher state of concern for the future, and concern for other beings. And even in 2006, nothing was more concerning to me than global heating.

Why most people can and do look away, and how to get them instead to pay urgent attention, might be the single most important question for humanity. The ability of the human mind to 'not think about' the climate and ecological emergency is the key to why it exists in the first place. If we all faced it, square on, together, we would surprise ourselves by how quickly we stopped it.

Before I go any further, I want to say that I currently find it incredibly hard to write about climate. I didn't always find it challenging; it used to be cathartic. I started writing my first book in 2012 and genuinely enjoyed the process. My ideas felt fresh. I had to learn a huge amount, and I enjoyed that feeling. Above all, I hoped that the book might make a difference. Now, a decade later, things are hotter than ever. In 2012 I thought nearly everyone would have switched out of denial and into climate emergency mode by now, at this intensity level of heat and disaster. I no longer feel that my writing makes much of a difference. Instead, it now mainly makes me feel as if I don't understand other people very well.

Before providing more of my personal perspective as a climate scientist and activist, here is a very brief overview of the planetary situation:[1]

**Fact cluster #1:** Pretty much anywhere you look scientifically within the Earth system, you see trends. Ocean heat is increasing. Surface air temperature

DOI: 10.4324/9781003436096-4

is increasing. Ice sheets and glaciers and sea ice are melting. Sea level is increasing. Ocean acidity is increasing. Atmospheric carbon dioxide, methane and nitrous oxide concentrations are increasing. Spring is coming sooner. Plant and animal ranges are shifting poleward. The frequency of extreme precipitation events is increasing. Extinction of species is rapidly accelerating. There are trends in the ocean, land, atmosphere, ice and soil; and in all things that creep, crawl, swim, fly and grow therein. On a planet in balance, all of these trends would instead be flat lines, some variation from year to year but on average, over multiple years, no trends. But instead, we see Earth system variables moving rapidly in one direction. Think of it like a giant monitoring board, with thousands of lights that should all be green but most of them are now flashing red and alarms are sounding. Taken together, these trends reveal that something is extremely wrong. Personally, I find them terrifying.

**Fact cluster #2:** All of these trends are due to humans. The two main causes are (1) dredging up, selling and burning staggering and still-increasing quantities of fossil fuels (coal, oil, fossil gas); and (2) growing, selling and slaughtering tens of billions of sentient beings ('livestock') every year. The fossil fuel industry is responsible for roughly 75–80% of global heating, and industrial animal agriculture is responsible for roughly 15%. These industries cause methane, carbon dioxide and nitrous oxide levels in the atmosphere to increase. These three molecules trap outgoing infrared energy, pushing Earth out of energy balance and forcing it to heat up.

**Fact cluster #3:** Many of the changes will be effectively irreversible on timescales that matter to humans and to human civilisation. This will be a hot planet for hundreds of years, and without aggressive mitigation, this will be a hot planet for thousands of years. Sea levels will be higher for thousands or tens of thousands of years. And biodiversity loss, a species-poor planet, will last for about ten million years.

**Fact cluster #4:** Billions of human lives are potentially at risk, from, amongst other things, extreme heat, famine or geopolitical breakdown and warfare due to intense pressures from multiple climate impacts. It still feels taboo to talk about civilisational collapse, and I think this explains why there has been so little discussion on this topic, let alone rigorous academic investigation. However, I think 'collapsology' is a shockingly understudied topic. Civilisation is delicate and vulnerable, especially in its current form with its vast, interconnected, brittle, complex and poorly understood supply chains. The climate forces pushing against the systems of civilisation – insurance, economics, agriculture, infrastructure, energy, transportation, politics and so on – are getting stronger by the day. The global poor are already experiencing the worst impacts, even though they have barely contributed to the emergency.

**Fact cluster #5:** Fossil fuel industry executives, lobbyists and flacks have been lying and colluding to systematically block action for decades – since at least the early 1990s despite having a surprisingly precise understanding of the planetary scale of the dangers as well as their irreversible and existential nature. It is less well-documented how long animal agriculture executives have known, but they certainly do now. Indeed, this irreversible damage is being perpetrated by a relatively small number of extremely rich people who are profiting from it and actively working to make it worse!

**Fact cluster #6:** It is not too late to fight hard, and it will never be too late to fight hard. Every tiny bit of fossil fuels burned makes the planet a little hotter, and every bit of additional heat makes Earth a little more unliveable for humans and other beings. We have *not* entered a 'new normal'. Instead, we are standing on an escalator taking us to ever higher levels of irreversible damage day by day and there is no plateau: the damage is a monotonic function of the greenhouse gases humanity emits, and that function is still poorly understood.

I have been aware of this basic story since 2006, although my understanding and perspective have deepened and changed over time. The psychological strain of holding this knowledge, while being surrounded for almost two decades by people who act as if nothing is wrong and who think I'm 'nuts', is immense. Cassandra Syndrome might be particularly challenging for scientists because we have such a depth of interlocking, detailed and absolutely compelling evidence supporting our view. People on Twitter say things like, 'What if Peter is right?' seemingly without realising that I can see it. A journalist interviewed me and asked if I were 'a sort of "climate Gandhi" or a madman'; I rejected his framing, because I'm neither – I'm just a person who sees a serious problem for what it is. When I hear planes flying overhead, part of me feels like screaming, but part of me has grown resigned, because almost no one else sees any problem with planes flying everywhere all the time during a climate emergency.

If we had started acting in emergency mode years ago, say in the 1970s, or even the 1990s, when the danger was already extremely well-understood scientifically, humanity could have easily avoided even a 1°C level of global heating. There would have been no serious impacts. As it is, we are on track to blow past 1.5°C in the early 2030s and the impacts are already severe – for example, some experts feel that the Amazon rainforest has already passed its tipping point – although still only a hint of where they could soon get. In my opinion, to sensibly risk-manage the possibility of widespread collapse, we should be decreasing emissions globally by *at least* 10% per year, starting *immediately*. This would amount to a shift into a climate emergency mode. However, we are still barely making a collective effort. Indeed, world leaders of all major nations, including the Democrats and the Biden administration in the US, are actually doing essentially everything they can to *expand* the fossil fuel industry. Meanwhile, heatwaves are off the charts, both on the land and in the ocean, ice melt appears to be accelerating beyond projections, and Earth's breakdown is generally stunning scientists with its rapidity.

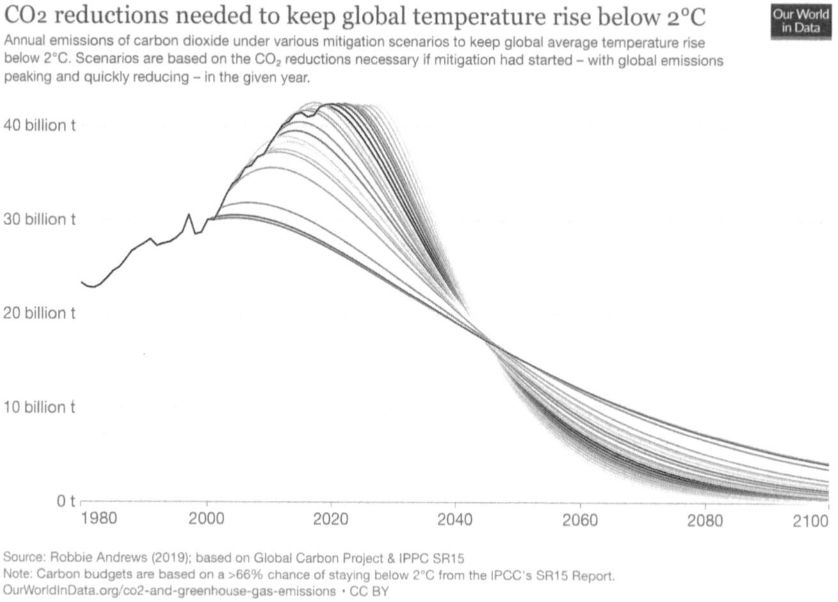


*Figure 1.1* The 'ski-slope' graph.[2]

The real question is this: If the stakes are so clear and the relevant information is so readily available, why is there still so little action to stop the damage? The White House in the US touts the Inflation Reduction Act as the largest climate bill in the history of the world by any country, but the spending earmarked for climate amounts to just 6% of what the US spends on its military. And we are still subsidising the fossil fuel industry to the tune of a trillion dollars a year.

Achieving 10% per year reductions will require thousands of emergency mode policies, at federal, state and local levels, designed to end the easy stuff immediately: private jets, private mega yachts, frequent flying, industrial agriculture and new fossil-fuelled power plants and pipelines; create and implement well-defined plans and milestones for the hard stuff – such as decarbonising transportation, buildings and agriculture; and assist rapid transformation in other countries. For these policies to work, they will need to protect the poor and middle class at the expense of the rich. This is not just for ethical reasons – it is a practical necessity. Otherwise, the policies will be unpopular, the poor and middle class will revolt against them, and they will be reversed through elections or other means. This same principle operates internationally: it is critical to get large Global South nations on board, otherwise the decarbonisation of rich Global North nations will not translate to a global end to emissions. However, modern democracies have been corrupted by corporations and the rich, who leverage their vast wealth to stop policies unfavourable to their profits – even at the cost of the liveable planet.

I have spoken to fossil fuel executives. They don't see an emergency, although they are certainly aware of climate change, and they don't take any responsibility. The Shell executives I spoke to, and I think this is probably typical, kept repeating that they would 'go as fast as our customers want'. They do take the profits though, and it bears repeating that they have done all they can to prevent action, and *continue* to do so, as was shockingly apparent in the US congressional questioning in 2021 of the CEOs of Exxon, Shell, Chevron and BP. Meanwhile, the rich don't want to be taxed at a higher rate or lose their outsized fossil fuel privilege in which the top 1% of emitters cause over ten times the climate impact of the average human. Simply put, the rich and the powerful want to maintain business as usual, the fossil-fuelled status quo, which benefits them personally in the short term, for as long as possible.

Therefore, rapid change will need to come from the grassroots climate movement. Once the movement is stronger than the fossil fuel industry, policies will come in quick succession, creating rapid systemic change. However, the movement has been extremely slow to grow. As I write this in 2023, it is still not possible to get large numbers at climate protests and actions. This is partly because the rich own the mainstream media, and they use it to nudge the public away from radical change and toward preserving the status quo. This is also partly why the media has until recently barely reported the climate emergency, and why even in 2023 it reports without appropriate urgency. The other reason why it is hard to get people to attend protests, and why the media will not report them with appropriate urgency, is because society remains in collective denial. There is a clear, imminent, well-studied and well-understood danger to civilisation, but society is still acting as if everything is normal.

The physical reality and scientific facts of Earth breakdown are legion, but our brains are more powerfully influenced by social norms and the postures and attitudes of the people around us. This leads to inaction due to socially constructed silence.

Because so few people are acting as if Earth breakdown is an emergency – not even some scientific experts and certainly not the media – the vast majority of people assume it is *not* an emergency. The few Cassandras are dismissed relatively easily. When young people, desperate to shake society out of this deadly torpor, do things such as throw soup at a painting, the people in power clutch their pearls and excoriate the activists in far stronger terms than they use to decry fossil fuel executives who are responsible for the ongoing irreversible breakdown of our entire planet, threatening billions of lives and life on Earth as we know it. They are in denial.

I have tried many things to help push society into climate emergency mode. I became a climate scientist. I have written articles and a book. I have co-created a climate app and a media non-profit organisation. I have given talks and helped organise protests. But nothing I've done has worked as well as climate disobedience. In my experience, arrestable civil disobedience is essentially a communications technology that breaks through the socially constructed silence by *demonstrating* emergency action.

I've been asked by many journalists – probably well over a hundred – 'What gives you hope?' My hope comes from the fact that humanity has barely started to try to stop Earth's breakdown, because there is still every possibility that if we actually did take strong concerted action, we would succeed. Earlier I mentioned that the new US climate bill sets aside an amount that comes out to about 6% of the military budget. Imagine what we could do if we spent, say 10% of what we spend on the military to preserve what's left of the liveability of our planet? Or 20%. Heck, even 50%. Or – hear me out – what if we humans actually put living in balance on Earth, the one place in the cosmos we know to have life, this absolutely wonderful spaceship in this crazy, cold, violent, baffling universe, as a *higher* priority than blowing each other up? That's what gives me hope. But maybe I'm being naïve. As a society, we seem to still be very intent on blowing each other up and to have precious little gratitude for this wonderful Earth.

For a person to act with appropriate urgency, they must be out of climate denial. In my experience, coming out of climate denial requires coming to terms with climate grief. We are experiencing tremendous loss on planet Earth, and when the fact of that tremendous loss is accepted, grief is a natural response. I believe that even most climate scientists are still avoiding facing this grief. It is certainly true that climate scientists do not talk about their grief with each other very often. However, when an appropriate, intentional space is created, a latent need seems to emerge. I have met with colleagues on several occasions specifically to talk about climate grief. Every time, after overcoming some initial awkwardness, it has turned out to be a powerfully cathartic experience for those involved.

I am routinely trolled by climate deniers. I have even received death threats. But the comments I find most hurtful tend to come from other climate activists who feel less urgency than I do. Once, I was told by such a person, in a mean-spirited way, to 'seek therapy'. I think this statement suggests that the person was still in denial: my psychological experience is a response to physical, permanent and dangerous changes occurring on Earth, which will continue to worsen whether or not I go to therapy. However, as it happens, I have been to therapy, and I think it is critical for those working in climate fields to also seek similar help, *and* for therapists to be prepared to deal with intense climate issues. As scientists, journalists, activists and others face and accept their climate grief, they will begin to act in emergency mode, and this will help break through the socially constructed silence and push society into emergency mode.

Many of us are living in the queasy dual reality of recognising that the Earth is breaking down while living in a society that ignores this breakdown.

I feel this every time I hear a plane fly overhead, for example. The climate activist community is in real need of support from climate therapists. We have been shouldering a very heavy load – some of us for decades. We are feeling gaslit. We cannot understand the lack of empathy, the denial and lack of action all around us. We need climate-aware therapy, and we need it now.

## Notes

1 For more on the science on climate change and its impacts see:
Wallace Wells, D. (2019) *The Uninhabitable Earth: A Story of the Future*. London: Allen Lane.
Recent IPCC reports, for example:
www.ipcc.ch/report/sixth-assessment-report-working-group-i/
www.ipcc.ch/report/sixth-assessment-report-working-group-ii/

2 https://ourworldindata.org/grapher/co2-mitigation-2c
https://creativecommons.org/about/cclicenses/
Also see author's article in *The Guardian*, 'Joe Biden Must Declare a Climate Emergency and He Must Do It Now'. Available at www.theguardian.com/commentisfree/2023/jul/27/joe-biden-climate-emergency-peter-kalmus [Accessed 27 June 2023].

Voice 2

# It's hot as fuck and I need to rest my eyes

*Fehinti Balogun*

DOI: 10.4324/9781003436096-5

*Plate 2.1*

This painting is called 'It's hot as fuck and I need to rest my eyes'. When I painted it, it was hot as fuck. Another heat wave. Climate anxiety was making me feel like I was being swallowed by tar. This is the only way I know how to climb out. I make art and hope it makes some kind of a difference.

Fehinti Balogun, UK is an actor who has worked in theatre, film and television. Alongside his acting career, he delivers talks on climate change aimed at creating more rounded inclusion in the conversation and has spoken at the UN COP 26 climate summit and other forums.

Chapter 2

# The Mental Health and Emotional Impacts of Climate Breakdown: Insights from Climate Psychology

*Judith Anderson and Rebecca Nestor*

## Introduction

Facing climate breakdown is a daunting prospect. Psychological professionals, with their skills in helping people face existential fears and navigate defences, are essential to the task. Although the focus of this book is the psychological and emotional impacts, they cannot be separated from impacts on the human body and on societal structures, which are set in the broader multiple destructions of planetary health that result from environmental degradation and biodiversity loss. The first section of the chapter may be a difficult read for those in the Global North, bringing us up against painful realities from which we are protected by privilege. The second section of the chapter explores how we defend ourselves against those realities in psychosocial terms.

## Facing Climate Breakdown

These two authors are both UK-based and have experienced heatwaves, floods, and storms, made more frequent by climate change. The impact on us has been slight but we both know others who have suffered significant disruption to their lives because of these events. Many readers have friends, family and colleagues in other countries who have been directly impacted by severe climate change related events such wildfires, storms, floods, and heat domes. Despite these connections that bring climate breakdown into our reality, it can be difficult to keep in mind even more serious disasters, such as in Pakistan through the autumn of 2022 when 33 million people were impacted by floods that left one-third of the country underwater. How can we find ways to process the unfolding catastrophe in the Horn of Africa where an unprecedented four-year drought has left an estimated 22 million at risk of starvation, a situation worsened by local conflict and the impact of the war in Ukraine on grain supplies?

The language used to refer to climate science has an impact on body and mind; how could it not? Phrases such as 'the 1.5 global climate target is on life

DOI: 10.4324/9781003436096-6

support', and 'we are getting dangerously close to the point of no return' (Guterres, 2022) evoke images of death and critical illness, reminding us of feared situations.

Working with University College London, *The Lancet* journal has published exhaustive details of the unprecedented threat that climate change poses to human health globally – in the medium term, as a danger to already vulnerable continents such as Africa and, in the long term, as a potential species-extinction process. Since the Paris Agreement came into force in 2016, the Lancet team has produced annual Countdown reports providing independent assessments of progress towards the goals of the Paris Agreement (Watts et al., 2017). They describe the impacts of rising sea levels and flooding with the prospect of death, injury, ill-health or disrupted livelihoods as a major concern, particularly given that 10% of the world's population live or work in low-lying coastal zones and island states.

Longer-term impacts of these events include the breakdown of infrastructure networks and critical services such as electricity, water supply and health and emergency services. We will witness and/or be caught up in:

- Higher mortality and morbidity during periods of extreme heat
- Food insecurity and the breakdown of food systems, particularly for poorer populations
- Pollution
- Vector-borne disease changes
- Ecosystem changes and loss of biodiversity
- Poverty, migration and conflict

This raises a further issue to take into psychological work in this field. Climate breakdown is a function not only of fossil fuel use but of structural inequality; 98% of the people currently severely affected by climate change and 99% of those dying because of climate change live in the lower income countries (Global Humanitarian Forum, 2009).

There are colonialist, extractive antecedents of the combined climate, biodiversity, and environmental crises. Relational Psychoanalyst Donna Orange in her book *Climate Crisis, Psychoanalysis and Radical Ethics* (Orange, 2016) gives a searing account of the extent to which our western blindness to climate change may well be because it happens more to people we have become encultured to ignore through our history of involvement in colonialism and slavery.

This unconscious bias plays out in all our environmental organisations. The environment and conservation professions are amongst the least diverse in the UK. The RACE Report (2022) highlights the need for greater representation in sustainability and climate action. Knowing this, we must practise inclusion from a position of recognising our unconscious bias. The psychotherapy profession is working to address issues of racism and unconscious bias in all aspects of its work (Ryde, 2009; Morgan, 2021; Ellis, 2022) and it is essential to include in any psychological work on climate breakdown.

## Temperature and Our Animal Body and Mind

As well as affecting our physical health, temperature affects our mental health. This impact is less well known and potentially dangerously neglected by psychological practitioners, despite its relevance for those working with mental health teams, with family, friends or neighbours who are vulnerable, or who have in their practices individuals with severe mental illness who may be on anti-psychotic medication.

In the UK each degree of increase in mean temperature above 18°C was associated with a 3.8% and 5.0% rise in suicide and violent suicide respectively (Page, Hajat and Kovats, 2007). Similar effects are seen globally when temperatures rise above that to which individuals are adapted.

Further, Page, Hajat and Kovats (2012) showed that patients with mental illness had an overall increase in risk of death of 4.9% per 1°C increase in temperature above the 93rd percentile of the annual temperature distribution. Younger patients and those with a primary diagnosis of substance misuse demonstrated the greatest mortality risk.

A recent study associated rises in ambient temperature with the prevalence of Intimate Partner Violence (Zhu et al., 2023) with a 1°C increase in the annual mean temperature associated with a 4.5% increase in IPV prevalence.

The increased risk of death during hot weather in patients with psychosis, dementia and substance misuse has implications for tailored public health strategies during heatwaves.

## Impact of Slow and Repeated Effects of Disruptive Climate Events

Research in Australia looked at the psychological impact of prolonged repeated droughts (O'Brien, 2014). The initial review of state-wide surveys of mental health seemed to show little change. The significance arose when the population was sorted into rural vs urban; the rural farming communities where the practical impacts were devastating were more likely to experience mental health impacts, not surprisingly more often in those who already had psychological problems and/or were vulnerable for other reasons. It was postulated that the urban group was distanced from the direct impact of droughts; dry and sunny days were experienced as a time to socialise outside.

The longer-term effects of weather-related hazards can be especially perilous, including food shortages, homelessness and displacement, damage to public infrastructure, power and connectivity, loss of agricultural land and sacred places. All of these can impair social cohesion, undermining crucial support for psychological well-being.

Research on floods in the UK in 2007 (Paranjothy et al., 2011) showed increased rates of depression and anxiety three to six months later, with the most vulnerable affected more (e.g., existing psychological problems, poverty, comorbid physical disabilities). In northern Europe, it is predicted that the greatest burden of disease of any kind caused by climate change will be adverse mental health problems following floods and storms (Menne and Murray, 2013). A number of factors have been found to increase vulnerability to these psychological impacts of extreme weather, including older age, pre-existing medical conditions, inadequate insurance cover and social deprivation (Paavola, 2017).

When these tangible losses are present in the consulting room, what is it like to sit with such traumatic events as part of an inevitably escalating pattern whose solutions have been ignored by governments?

What might that mean for practitioners themselves caught up in such events? In the aftermath of Hurricane Katrina, many therapists left New Orleans never to return. Those that did resumed their practices in adverse circumstances, their own homes often leaking. One analyst writes of a change of frame as they lived through the mopping-up exercise. Tradespeople were in such short supply that it became quite ordinary and understood for both client and therapist to keep their phone on in case a plumber called (Boulanger, 2013).

What therapists had to learn from COVID, when we were all caught up in our worlds being changed, was how to hold a balance of acknowledging overwhelmingly shared experiences (not to do so would make therapy a kind of delusional, 'let's pretend' space), at the same time holding the space for difference in experience and the experience of that difference.

## Climate Emotions and Defences Against Them

Climate psychology has introduced new understandings of the complexity of climate emotions. In this section of the chapter, we explore some of these, with a focus on how people – particularly in the Global North – may need to defend collectively against these emotions through disavowal and other forms of splitting. Grief and loss are at the centre of these emotions (Randall, 2009; Willox, 2012; Lertzman, 2015; Head, 2016; Randall and Hoggett, 2019). Global North social structures and education encourage us to deny our dependence on each other and on other creatures and systems in our shared home (Rust, 2007; Randall, 2013). The reality of human dependence evokes early feelings of helplessness and activates early defences such as splitting and projection, pre-Oedipal narcissistic rage and omnipotence, and Oedipal and depressive guilt (Randall, 2013): we experience both love and hate for our Mother Earth, but we are not supported to acknowledge this ambivalence.

Instead, we are socialised into believing in the myth of our independence, and in the right of the wealthiest humans to exploit both the planet and poorer, typically black and brown humans (Rust, 2007:4). We disconnect from the intimacy (Rust, 2007) and creatureliness (Hoggett, 2020) that remind us that we are inescapably in relationship with, interdependent with, the more-than-human world.

As a result, many of us are not in touch with the climate/eco-grief that arises from the long, slow loss of our fellow creatures. Given that climate grief is not the uncomplicated grief that arises from a single loss, it will continue to be with us for the rest of our lives and must become part of the fabric of life to be inhabited, or perhaps to inhabit us.

Explored in detail by Pihkala in Chapter 12 of this book, grieving and mourning are an important aspect of healing the disconnect from ourselves and from the natural world, and of recognising animal and plant deaths as 'grievable' (Randall, 2009; Willox, 2012; Lertzman, 2015). But in the absence of support for grieving, climate psychologists have noticed defensive responses to these often unbearable feelings, including 'socially constructed silence' (Norgaard, 2011, 2018; Hoggett and Randall, 2016); apathy (Lertzman, 2015), and dependence on phantasy saviours such as technology and individual leaders (Randall, 2009; Randall and Brown, 2015). Various kinds of defensive insistence show up: insistence on optimism and positivity (le Goff, 2017), insistence on the certainty of collapse or apocalypse (Hoggett, 2011) and insistence on the role of emerging technologies in fixing the problem and enabling emissions to continue (Long, 2015).

For those in the Global North, an emerging awareness of our fossil-fuelled exploitative history can create strong pressures to feel personal guilt (Randall, 2013). Climate psychologists link guilt to the moral injury presented by climate change (Weintrobe, 2020; Hickman and Marks et al., 2021) and these difficult feelings may evoke the primitive responses of envy and shame (Orange, 2016). Other writers point to the multiplicity and messiness ('biodiversity') of the emotional impact (Hickman, 2020; Stanley et al., 2021; Pihkala, 2022) and discuss other emotions including rage and terror.

These psychological processes operate against a backdrop of social and political dynamics which often intensify them. For example, Kari Norgaard's (2011) study of 'socially constructed silence' on climate change in a Norwegian community provides a convincing illustration of how defence mechanisms work at the social level in a context of acute contradiction between, on the one hand, social and national identities that include environmental concern, and on the other the reality of a society entirely dependent on fossil fuels. Sally Weintrobe (2021) makes a convincing case that the structures of late capitalism promote the uncaring parts of ourselves at the expense of the caring parts. As she argues, advertising encourages us to believe that we are excepted from the demands of reality (Weintrobe, 2020, 2021), that we must have whatever we want because, as Rosemary Randall

(2005: 172) puts it, 'desire or demand has been translated into need'. These are the perverse dynamics proposed by Susan Long in a 2015 paper on fossil fuel companies: omnipotence and certainty, violent attacks on the object of dependence and abusive behaviour.

## Eco-distress

As people awaken to the reality that has previously been disavowed or denied outright, they will begin to experience distress. The familiar term 'eco-anxiety' is a manifestation of this awakening. Some practitioners now use the term 'eco-distress' to acknowledge the wide variety ('biodiversity') of emotional responses, which we touch on above. Solastalgia is a specific eco-emotion defined by Glenn Albrecht (2019) as 'the homesickness you have when you are still at home', when the lived environment is changing in ways that are distressing. In many cases this is in reference to global climate change but can also occur following more localised situations such as volcanic eruptions, drought, major landslides and destructive mining techniques.

The American Psychological Association report, Mental Health and Our Changing Climate: Impacts, Implications, and Guidance states:

> We can say that a significant proportion of people are experiencing stress and worry about the potential impacts of climate change, and that the level of worry is almost certainly increasing.
>
> (Clayton et al., 2017)

Using the then more familiar term 'eco-anxiety' they describe it as 'watching the slow and seemingly irrevocable impacts of climate change unfold, and worrying about the future for oneself, children, and later generations … being deeply affected by feelings of loss, helplessness and frustration due to an inability to feel like they are making a difference in stopping climate change'.

Some scholars differentiate between climate worry or concern on the one hand, and anxiety on the other. Clayton and colleagues developed a measure of climate anxiety, which they tested on US adults – finding that anxiety was not uncommon, especially among young adults (Clayton and Karazsia, 2020). A global study of 10,000 young people in ten different countries (Hickman et al., 2021) found that respondents across all countries were worried about climate change (59% were very or extremely worried and 84% were at least moderately worried). Similarly, a recent UK study (Whitmarsh et al., 2022) found three-quarters of participants reported being worried about climate change. However, the same study used the Clayton measure of climate anxiety itself and found overall scores were far lower, with only 4% registering moderate or severe climate anxiety. As the methodology did not collect experiences of other climate emotions including fear, anger, grief, despair, guilt and shame, the authors note the possible underestimation of broader emotional impacts (and see Coffey et al., 2021; Pihkala 2020, 2022; Soutar and Wand, 2022). More than 50% in the Hickman study reported feeling sad, anxious, angry, powerless, helpless and guilty, with more than 45% of respondents saying that their feelings about climate change negatively affected their daily life and functioning.

It is important to reiterate that these are not pathological states but responses to reality. This is widely recognised across the spectrum of psychological professions. The Royal College of Psychiatrists website (RCPsych, 2022) publishes helpful fact sheets for young people and their parents/carers about eco-distress, emphasising that these are normal, understandable reactions. Accordingly, there has been a reluctance to include 'eco-anxiety' or 'eco-distress' in diagnostic manuals.

The fact that eco-distress is 'normal' and understandable does not preclude the need for psychologically thoughtful interventions, both at an individual level and/or within a community (see Xue et al., 2023 and Baudon and Jachens, 2022 for overviews of these). We understand this in relation to 'normal' grieving. Additionally, understandable climate emotions can coexist with other states.

Those who are most vulnerable seem in our experience to be those who work, learn or live in situations where the facts are not concealed. In our clinical and organisational work, we find scientists, civil servants in the environmental sectors, NGO employees and volunteers, teachers, pupils and students and activists all describing a wide range of difficult emotions connected to their work, interests and concerns. Many feel troubled, anxious and sometimes terrified at the prospect of having children.

## Terror Management Theory, Climate Trauma and Climate Change as a Super-wicked Problem and a Hyperobject

A field adjacent to climate psychology, Terror Management Theory (TMT) proposes that the knowledge of our own death creates anxiety and defensive responses, such as difficulties with identity and a raised concern for social status. Climate change is increasing the salience of mortality in our minds and expanding 'ordinary' awareness of death to encompass awareness of the potential for our own extinction and the deaths of other species. As a result, we are seeing increasing polarisation in society in the Global North, as part of which extreme positions on the climate are becoming more evident. Some TMT scholars (Adams, 2016; Stolorow, 2020, 2021; Bermudez, 2021) are proposing that in a self-perpetuating dynamic, forced migration and displacement of human and animal communities (which is both an immediate impact of fossil fuel extraction and one of the consequences of climate change) create unacknowledged anxiety which leads to forms of defence such as barriers to free movement, the creation of billionaires' bunkers in New Zealand and further denial of responsibility (Bettini, 2019; Whyte, 2020).

Climate denial is becoming less widespread but more extreme (Hollway, 2020). Through the use of what William Lamb and colleagues call 'discourses of climate delay' the denialists are also borrowing (Lamb et al., 2020) from its opposite, climate apocalypticism (Hoggett, 2011), so that 'it's too late' has replaced 'it's not happening' as a reason to keep the status quo. Splitting is evident in governments' grand targets and public statements of the severity of the crisis that appear alongside risibly tiny practical commitments.

In this context, gestalt therapist Steffi Bednarek (2021) builds on ecopsychologist Zhiwa Woodbury's (2019) taxonomy of climate trauma to paint a compelling picture of the traumatising impact of the extractivist activities of some groups of humans, over generations, on all species and the natural world, including humans. The subject of collective climate trauma has also been explored from the perspectives of philosophy, humanities, cultural studies, and sustainability (Kaplan, 2015; Doppelt, 2017; Richardson, 2018; Zimmerman, 2020).

Systemically, climate change has been seen as a 'super-wicked' problem (Lazarus, 2008; Levin et al., 2009, 2012) – defined as one that will become more difficult to address the longer we leave it; in which the most power to act is held by those with the least immediate incentive to do so; and where there is no authority with the global reach needed to act. This absence of authority carries a significant emotional impact through reminding us of the global scale of the problem some humans have created, and the smallness of all humans in relation to it. A similar emotional impact is also present in the philosophical concept of climate change as a hyper-object, defined by philosopher Timothy Morton (interviewed in this book as Voice 4) as 'things that are massively distributed in time and space relative to humans' (Morton, 2013: 1) – put simply, things that are too big for us to cope with.

If the climate crisis is too big for us, perhaps we need to reconnect with being small – even, as Paul Hoggett puts it, to 'embrace our smallness' (Hoggett, 2023). Who better than the therapeutic professions to support this reconnection?

## References

Adams, M. (2016) *Ecological Crisis, Sustainability and the Psychosocial Subject: Beyond Behaviour Change*. London: Macmillan.

Albrecht G. A. (2019) *Earth Emotions. New Words for a New World*. Ithaca: Cornell University Press.

Baudon, P. and Jachens, L. (2021) A Scoping Review of Interventions for the Treatment of Eco-anxiety. *International Journal of Environmental Research and Public Health*, 18(18): 9636. doi 10.3390/ijerph18189636.

Bednarek, S. (2021) Climate Change, Fragmentation and Collective Trauma: Bridging the Divided Stories We Live By. *Journal of Social Work Practice*, 35: 5–17.

Bermudez, G. (2021) *Walls Against Nature? Social Defense Systems, Climate Change, and Eco-anxiety. The Walls Within: Working with Defenses Against Otherness.* Annual Meeting of the International Society for the Psychoanalytic Study of Organizations. Berlin.

Bettini, G. (2019) And Yet it Moves! (Climate) Migration as a Symptom in the Anthropocene. *Mobilities*, 14: 336–350.

Boulanger, G. (2013) Fearful Symmetry: Shared Trauma in New Orleans After Hurricane Katrina. *Psychoanalytic Dialogues*, 23: 31–44.

Clayton, S. and Karazsia, B. T. (2020) Development and Validation of a Measure of Climate Change Anxiety. *Journal of Environmental Psychology*, 69: 101434.

Clayton, S., Manning, C. M., Krygsman, K. and Speiser, M. (2017) *Mental Health and Our Changing Climate: Impacts, Implications, and Guidance*. Washington, DC: American Psychological Association, and ecoAmerica.

Coffey, Y., Bhullar, Y.N., Durkin, J., Islam, M. S. and Usher, K. (2021) Understanding Eco-anxiety: A Systematic Scoping Review of Current Literature and Identified Knowledge Gaps. *The Journal of Climate Change and Health*, 3: 100047.

Doppelt, B. (2017) *Transformational Resilience: How Building Human Resilience to Climate Disruption Can Safeguard Society and Increase Wellbeing*. London: Routledge.

Ellis, E. (2022) *The Race Conversation*. UK: Confer Ltd.

Global Humanitarian Forum, (2009) *Human Impact Report: Climate Change – The Anatomy of a Silent Crisis*. Geneva.

Guterres, A. (2022) *from Secretary-General's remarks to High-Level opening of COP27*. Available at www.un.org/sg/en/content/sg/speeches/2022-11-07/secretary-generals-remarks-high-level-opening-of-cop27#:~:text=The%20science%20is%20clear%3A%20any,the%20point%20of%20no%20return.

Head, L. (2016) *Hope and Grief in the Anthropocene*, Abingdon: Routledge.

Hickman, C. (2020) We Need to (Find a Way to) Talk About … Eco-anxiety. *Journal of Social Work Practice*, 34: 411–424.

Hickman, C., Marks, E., Pihkala, P. et al. (2021) Climate Anxiety in Children and Young People and Their Beliefs About Government Responses to Climate Change: A Global Survey. *Lancet Planetary Health*, 5: e863–e873. doi 10.1016/S2542-5196(21)00278-3.

Hoggett, P. (2011) Climate Change and the Apocalyptic Imagination. In: H. Brunning (ed.) *Psychoanalytic Reflections on a Changing World*. London: Karnac Books.

Hoggett, P. (ed.) (2019) *Climate Psychology: On Indifference to Disaster*. London: Palgrave Macmillan.

Hoggett, P. (2020) The Nature Within. *Journal of Social Work Practice*, 34: 367–376.

Hoggett, P. (2023) Imagining Our Way in the Anthropocene. *Organisational and Social Dynamics*, 23(1): 1–14.

Hoggett, P. and Randall, R. (2016) Socially Constructed Silence? Protecting Policymakers from the Unthinkable [Online]. Available: www.opendemocracy.net/transformation/paul-hoggett-rosemary-randall/socially-constructed-silence-protecting-policymakers-fr [Accessed 14 August 2021].

Hollway, W. (2020) CPA Newsletter November 2020 – Climate CRISIS DIGEST: Petromasculinity [Online]. Climate Psychology Alliance. Available: https://mailchi.

mp/96dfdf1eb76c/a-tale-of-two-tipping-points-4716934?e=827f3c2c23 [Accessed 14 August 2021].

Hwong A. R., Wang, M., Khan, H. et al. (2022) Climate Change and Mental Health Research Methods, Gaps, and Priorities: A Scoping Review. *Lancet Planetary Health*, 6(3): 281.

Kaplan, E. A. (2015) *Climate Trauma*. Ithaca, NY: Rutgers University Press.

Lamb, W. F., Mattioli, G., Levi, S., Roberts, J. T., Capstick, S., Creutzig, F., Minx, J. C., Müller-Hansen, F., Culhane, T. and Steinberger, J. K. (2020) Discourses of Climate Delay. *Global Sustainability*, 3: e17.

Lancet Countdown (2015) www.thelancet.com/countdown-health-climate/about

Lazarus, R. J. (2008) Super Wicked Problems and Climate Change: Restraining the Present to Liberate the Future. *Cornell Law Review*, 94: 1153.

Le Goff, J. (2017) Des effets des discours positifs sur les angoisses liées au changement climatique. *Nouvelle Revue de Psychosociologie*, 2: 145–156.

Lertzman, R. (2015) *Environmental Melancholia: Psychoanalytic Dimensions of Engagement*. Hove, East Sussex: Routledge.

Levin, K., Cashore., B., Bernstein., S. and Auld, G. (2009) Playing it Forward: Path Dependency, Progressive Incrementalism, and the 'Super Wicked' Problem of Global Climate Change. *Earth and Environmental Science,* 6(50). doi:10.1088/1755-1307/6/50/502002. Bristol: IOP Publishing.

Levin, K., Cashore, B., Bernstein, S. and Auld, G (2012) Overcoming the Tragedy of Super Wicked Problems: Constraining Our Future Selves to Ameliorate Global Climate Change. *Policy Sciences*, 45: 123–152.

Long, S. (2015) Turning a Blind Eye to Climate Change. *Organisational & Social Dynamics*, 15: 248–262.

Menne, B. and Murray, V. (2013) *Floods in the WHO European Region: Health Effects and Their Prevention*. Denmark: WHO Europe. Available at https://iris.who.int/bitstream/handle/10665/108625/9789289000116-eng.pdf?sequence=1

Morgan, H. (2021) *The Work of Whiteness: A Psychoanalytic Perspective*. Abingdon: Routledge.

Morton, T. (2013) *Hyperobject*. Minneapolis: University of Minnesota Press.

Norgaard, K. (2011). *Living in Denial: Climate Change, Emotions and Everyday Life*. Cambridge, MA: MIT Press.

Norgaard, K. M. (2018). The Sociological Imagination in a Time of Climate Change. *Global and Planetary Change*, 163: 171–176. doi: 10.1016/j.gloplacha.2017.09.018.

Orange, D. (2016) *Climate Crisis, Psychoanalysis, and Radical Ethics*. London: Taylor and Francis.

O'Brien, L. V., Berry, H. L., Coleman, C. and Hanigan, I. C. (2014) Drought as a Mental Health Exposure. *Journal of Environmental Research*, 131: 181–187.

Paavola, J. (2017) Health Impacts of Climate Change and Health and Social Inequalities in the UK. *Environmental Health*, 16: 11.

Page, L. A., Hajat, S. and Kovats, R. S. (2007) Relationship Between Daily Suicide Counts and Temperature in England and Wales. *British Journal of Psychiatry*, 191: 106–112.

Page, L. A., Hajat, S., Kovats, R. S. and Howard, L. M. (2012) Temperature-related Deaths in People with Psychosis, Dementia and Substance Misuse. *British Journal of Psychiatry*, 200: 485–490.

Paranjothy, S., Gallacher, J., Amiot, R. et al. (2011) Psychosocial Impact of the Summer 2007 Floods in England. *BMC Public Health* 11: 145. Available at. https://bmcpublichealth.biomedcentral.com/articles/10.1186/1471-2458-11-145.

Pihkala, P. (2020) Anxiety and the Ecological Crisis: An Analysis of Eco-Anxiety and Climate Anxiety. *Sustainability* [Online]: 12.

Pihkala, P. (2022) Toward a Taxonomy of Climate Emotions. *Frontiers in Climate*, 3. www.race-report.uk [Accessed 30 April 2022].

Randall, R. (2005) A New Climate for Psychotherapy? *Psychotherapy and Politics International*, 3: 165–179.

Randall, R. (2009) Loss and Climate Change: The Cost of Parallel Narratives. *Ecopsychology*, 1: 118–129.

Randall, R. (2013) Great Expectations: The Psychodynamics of Ecological Debt. In: S. Weintrobe (ed.) *Engaging with Climate Change: Psychoanalytic and Interdisciplinary Perspectives*. London: Routledge.

Randall, R. and Brown, A. (2015) *In Time for Tomorrow? The Carbon Conversations Handbook*. Stirling: The Surefoot Effect.

Randall, R. and Hoggett, P. (2019) Engaging with Climate Change: Comparing the Cultures of Science and Activism. In: P. Hoggett (ed.) *Climate Psychology: On Indifference to Disaster*. Cham, Switzerland: Palgrave MacMillan.

RCPsych (2022) Eco Distress for Children and Young People. Available at www.rcpsych.ac.uk/mental-health/parents-and-young-people/young-people/eco-distress-for-young-people [Accessed August 2022].

Richardson, M. (2018) Climate Trauma, Or the Affects of the Catastrophe to Come. *Environmental Humanities*, 10: 1–19.

Rust, M-J. (2007) *Climate on the Couch: Unconscious Processes in Relation to Our Environmental Crisis*. London: Annual Lecture for Guild of Psychotherapists.

Ryde, J. (2009) *Being White in the Helping Professions: Developing Effective Intercultural Awareness*. London: Jessica Kingsley Publishers.

Soutar, C. and Wand, A. P. F. (2022) Understanding the Spectrum of Anxiety Responses to Climate Change: A Systematic Review of the Qualitative Literature. *International Journal of Environmental Research and Public Health*, 19(2).

Stanley, S. K., Hogg, T. L., Leviston, Z. and Walker, I. (2021) From Anger to Action: Differential Impacts of Eco-anxiety, Eco-depression, and Eco-anger on Climate Action and Wellbeing. *The Journal of Climate Change and Health*, 1: 100003.

Stolorow, R. D. (2020). Planet Earth: Crumbling Metaphysical Illusion. *American Imago*, 77: 105–107.

Stolorow, R. D. (2021) Heidegger's Angst and Apocalyptic Anxiety. *Metalepsis: Journal of the American Board and Academy of Psychoanalysis*, 1: 120.

Watts, N., Adger, W. N., Ayeb-Karlsson, S. et al. (2017) The Lancet Countdown: Tracking Progress on Health and Climate Change. *Lancet*, 389: 1151–1164.

Weintrobe, S. (2013a) The Difficult Problem of Anxiety in Thinking About Climate Change. In: S. Weintrobe (ed.) *Engaging with Climate Change: Psychoanalytic and Interdisciplinary Perspectives*. London: Routledge.

Weintrobe, S. (2013b) *Engaging with Climate Change: Psychoanalytic and Interdisciplinary Perspectives*. Hove: Routledge.

Weintrobe, S. (2020) Moral Injury, the Culture of Uncare and the Climate Bubble. *Journal of Social Work Practice*, 34: 351–362.

Weintrobe, S. (2021) *Psychological Roots of the Climate Crisis: Neoliberal Exceptionalism and the Culture of Uncare*. London: Bloomsbury.
Weller, F. (2015) *The Wild Edge of Sorrow: Rituals of Renewal and the Sacred Work of Grief*. Berkeley, CA: North Atlantic Books.
Whitmarsh, L., Player, L., Jiongco, A. et al. (2022) Climate Anxiety: What Predicts It and How Is It Related to Climate Action? *Journal of Environmental Psychology*, 83: 101866. doi: 10.1016/j.jenvp.2022.101866.
Whyte, K. (2020) Too Late for Indigenous Climate Justice: Ecological and Relational Tipping Points. *WIREs Climate Change*, 11: e603.
Willox, A. C. (2012) Climate Change as the Work of Mourning. *Ethics & the Environment*, 17: 137–164.
Woodbury, Z (2019) Climate Trauma: Toward a New Taxonomy of Trauma. *Ecopsychology*, 11: 1–8.
Xue, S., Massazza, A., Akhter-Khan, S.C., Wray, B., Husain, M.I. and Lawrance, E. (2023) Mental Health and Psychosocial Interventions in the Context of Climate Change: A Scoping Review. OSF Preprints. Available at: https://osf.io/cvthu/
Yixiang Zhu, M. S., He, C., Bell, M. et al. (2023) Association of Ambient Temperature with the Prevalence of Intimate Partner Violence among Partnered Women in Low- and Middle-income South Asian Countries. *JAMA Psychiatry*. Published online 28 June 2023. doi:10.1001/jamapsychiatry.
Zimmermann, L. (2020) *Trauma and the Discourse of Climate Change: Literature, Psychoanalysis, and Denial*. New York: Routledge.

Voice 3

# The heartbreak of rivers barely flowing

*Shelot Masithi*

The psychology of not having enough water to drink is immediate – this has heightened widespread anxiety and frustration amongst people, especially youths. As a young person, I am torn between solving the climate issues I live with daily in South Africa and being a youth free of this burden. World leaders pledged, to the whole world during COP26, that they'll put an end to climate inaction. However, it is frightening that we have not seen them since. Suffocating in thirst, eco-anxiety and the trauma left by climate catastrophes, young people like me continue to be the bigger person in the room.

Individual psychology is a largely British import into the African continent. Psycho-social Studies and Climate Psychology have the intention of shedding or at least reforming that legacy by paying respectful attention to the more collective traditions of the colonised world. If we took the perspective of *ubuntu* and applied it to water scarcity and the collective experience of thirst, where will that lead a community like my own to go beyond post-COP inaction and take into our own hands the prospect of improving that frightening future?

When I was growing up, I had the privilege of playing in the mountains and rivers and waterfalls. In different villages, growing up, I witnessed nature in its glorious form. One of these villages is where my great-grandmother and my grandmother lived. It had the most beautiful stream carrying water from the mountain to the village and farms, turning into one of the main rivers in my home region. That is where I learned to swim. It is so sad that when I go back to this village today, my heart breaks to see that river barely flowing.

Five years ago, it dried up completely and although the rains helped it back a little, it's still barely flowing.

## Experiencing water scarcity as a schoolgirl

Where I lived next, in my mid childhood, it was quite dry, just a rural village, no local industry. Back then from 2012 until 2017, we experienced severe water shortages that continue today. I was always worried about water,

DOI: 10.4324/9781003436096-7

whenever I left my house to go to school and return: what if tomorrow, there won't be water, do we have enough buckets to store water for ourselves when it runs out? In 2016 we experienced a water shortage for a month. The buckets that we had were not enough and ran out mid-month and the month dragged. We did not have enough money to buy water from those who had spare.

I remember one day, when in the morning I was ready for school, I stood in the middle of my room and wondered if I should take a bucket so that, on the way home, I could fetch water either from school or from the river. But I also thought if I take a bucket to school to fetch water on my way home, the other kids might bully me. I didn't think for a moment that they could also be experiencing the fear and anxiety of not having water. I didn't take the bucket. I just stumped out of my room and went to school. It was one of the most frustrating months in my grade school years. I couldn't even study properly because I was constantly thinking about water. And I hardly even knew about what other villages or other countries were going through when it comes to water and climate change.

## Realising the link with climate change

One day it happened that our geography teacher was teaching about climate change. However, I still didn't think that our water scarcity was due to climate change because he spoke of it as a crisis that's going to happen in the far distant future. But I started reading about it. As I read on, I realised that climate change is a real issue, but I was thinking about it happening in other places. I overlooked what was going on in my community.

## Community intervention

In my village, as we continued to experience water shortages, my local chief called a community meeting. It was June 2017. The idea was to source water from the mountain to provide running water to every house. That became a reality. We didn't have to worry about water again until October 2020. Then people started realising that there's no longer enough water in the mountains because it doesn't rain enough in the rainy seasons and because the village is expanding.

Now, because I have seen this in my community, and because I understand the links of climate change and water scarcity or drought, I also understand the need for psychosocial intervention. People had come together to solve one problem with a common goal. Like everywhere in the world, they needed access to water, for drinking, cooking, washing and growing crops. It is this unity that is needed. I realised that we have separated off climate change and issues like water scarcity as scientific problems and we ignore the social and psychological effects.

## Ubuntu

I learned about *Ubuntu*. *Ubuntu* means 'I am because we are'; it teaches that humanity is inextricably bound up with others; it emphasises community. And when I look at climate change and water scarcity, I think we need to look at this as the problem of the collective. When we come together, we form a collective hub that provides different knowledge, skills and expertise, as well as support. Water is a collective problem; climate change is a collective problem. When we try to solve them individually, we burn out easily. We need also to pay attention to the mental health, the psychological effects, of this problem that affects not only individuals but also the collective.

To solve these problems across a huge range of diverse communities is complex and cannot be achieved individually. So, when we try to separate ourselves from the community, pursue our own advantage, the problem remains. What is it that makes us so scared to build an inclusive community with no colour, conducive for every perspective?

I understand that most of you do not know what it feels like to be thirsty. I read that 75% of consumers in Europe pre-rinse their dishes! I also learned from the IPCC report of projections that about 700 million people in Africa could be displaced due to drought. What does that tell us? Must we continue to act individualistically?

Since COP26, our leaders are nowhere to be found with all the promises they made.

What do we do with this information? What have we really learned if we're still doing things individually? Let us question ourselves and our ways of doing things. Let us change our focus from the individualistic and explore the collective. What can it give us? Where can it take us?

Shelot Masithi, Activist, South Africa

Chapter 3

# Revisiting Ethics in the Context of Climate Breakdown

*Jenny O'Gorman*

## Introduction

This chapter considers ethics within the context of climate breakdown, exploring ways in which the ecological emergency challenges existing ethical frameworks, and the changing personal and professional demands on us as therapists.

Donna Orange proposes that our moral and ethical principles are grounded in preserving the social order, so we are now in need of radical ethics in order to give voice to that which has been silenced (2019). It is fitting, therefore, in the current context of the climate crisis to revisit what it means for us to practise ethically, including reflecting on that which has been silenced socio-politically, as well as within ourselves.

We are living through an emergent situation that impacts us all, so this work requires engagement with our own knowledge, understanding and emotional connection to the issues of environmental destruction in order to practise 'within the limits' of our skills (Health and Care Professions Council [HCPC] clause 3, 2016).

As therapists, we know the value of witnessing, of having all parts of the self recognised and acknowledged (Mighetto, 2022). When the psychosocial context of therapy is not acknowledged within the therapy room, the client may wonder if there can be space for their feelings around this.

Avoiding engagement with a client's broader psychosocial context could be interpreted as contravening the ethical tenet that therapists 'not harm or collude in the harming of clients' (UKCP clause 6, 2019). Giving the client permission to explore their thoughts and feelings about climate and ecological breakdown offers a form of deliberate witnessing that can validate the client's experience of the violence inherent in environmental destruction, directly, or through a form of moral injury (Hickman and Marks et al., 2021). As Weingarten (2003) puts it, this kind of deliberate or intentional witnessing 'has the potential for transforming violence at every level, from the personal to the societal'.

DOI: 10.4324/9781003436096-8

Engaging ethically whilst living through a time of ecological destruction requires thinking about specific elements of practice, such as disclosure and risk, as well as what it means for us as practitioners to advocate for our clients' – and our own – best interests. I introduce some of the key impacts of this on our practice below.

## Therapist Disclosure

One client[1] noticed and commented on an Extinction Rebellion (XR) badge on my mobility aid during an in-person session. When I confirmed that the symbol referred to XR, the client visibly relaxed. 'I talk about the climate sometimes,' they said, in a low tone bordering on confessional. 'Most people don't want to know. I see them roll their eyes … the part of me that cares isn't welcome in most of my circles'. They described how 'everything goes awkward. It's like people think I bring the environment up as a way of punishing them, or criticising how they live their own lives'.

A common facet of ethical frameworks produced by professional regulatory bodies is that the client should remain the focus of the work, their welfare and interests being of foremost importance. One interpretation of this suggests that practitioners should avoid self-disclosure wherever possible; but there will be inevitable and unavoidable, sometimes unconscious, self-disclosures, such as cultural dress, the presence of a wedding ring, or inferences made based on accent or even decorative choices within the therapy room (Gutheil, 2010). There are also times when it may be appropriate to self-disclose, as in the example above.

The question of what it means to be client-led and how much of the therapist's own identity can or should come into the work has been extensively interrogated (Barnett, 2011). There have been shifts in many modalities towards a more relational way of working, sometimes conceptualised as a form of cross-fertilisation allowing ideas and practices from across modalities to integrate (Hargaden, 2015). In this way, therapeutic practice is modelling the introjection of the wider world, introducing a broader dynamic which offers clients the opportunity to move beyond the intrapsychic and into the psychosocial. It is this extension into the psychosocial which is so important, providing the context for our clients' material, as well as for our own responses.

In my own practice, this is a theme which has emerged with other clients when thinking, speaking or acting on climate justice. Clients describe awareness of external pressures implicitly and explicitly encouraging them to stay silent, to suppress the part of the self which is interested, engaged and passionate. This could be considered to be part of a socially constructed silence, a form of projective shame (Lewin, 2020), one of the many ways in which perhaps guilt can be weaponised. As therapists, we must consider whether it is ethically sound for us to collude with such silence.

## Radical Actions

In October 2021, five young people working with the Sunrise Movement began a hunger strike outside the White House, pushing for a bill to be passed that would encourage utility companies to use renewables through both incentives and penalties. In the UK, activists have glued themselves to infrastructure and disrupted national sporting events, highlighting the absurdity of a 'business as usual' approach when significant environmental disruption is already occurring around the world. In Mexico, an Indigenous-led coalition occupied and subsequently took control of a factory that was extracting much-needed groundwater, subsequently re-launching the space as a community centre.

People have vastly different ideas around what constitutes a radical action. To some, throwing soup at the glass covering a painting is radical; to others, the same action would seem banal, an effective way of highlighting the absurdity of what activates our care and outrage, and what we fail to react to. UN Secretary-General António Guterres has said that 'Climate activists are sometimes depicted as dangerous radicals. But the truly dangerous radicals are the countries that are increasing the production of fossil fuels' (Guterres, 2022).

For therapists, taking radical action may involve acts of civil disobedience, but it may also mean challenging the status quo assumptions we have internalised. It could involve lobbying employers or therapy institutions to divest pension funds from polluting and extractive industries, or choosing ethical banking options and being explicit with clients about this choice. Radical action might include lodging motions at the AGMs of regulatory bodies to push for change within these organisations, or setting up co-operatives to share therapy rooms.

We are increasingly likely to be approached by clients who have taken part in radical actions, or the families of those who have. Providing such support often means working with people experiencing high levels of distress and navigating the changing boundaries between what is legal and illegal behaviour and what are moral and immoral actions.

## Risk and Safeguarding

In a recent session, a young client asked me: 'Why do people care when I self-harm?' They elaborated: 'Every time you get into a car it's self-harm. Our bodies are nature, so destroying nature is self-harm. It's hard to see why some people care about one but not the other'. Other therapists report clients fantasising about killing their children; therapeutically we could see this as an imaginal route to saving them from the suffering of living through impending systemic collapse.

Practitioners are expected to 'take immediate action to prevent or limit any harm' to clients (Feltham, Winter and Hanley, 2017). Whilst this is positioned as relating to harm caused by problems arising in the consulting room or linked to breakdowns within an organisation providing therapeutic services, it could be argued that an ethical position also involves taking action to limit harms likely to occur as a result of environmental breakdown in terms of the broader psychosocial environment, not just that of the therapeutic practice.

Common to all ethical frameworks is that practitioners should act in the 'best interests' of the client (BPC clause 1, 2011; UKCP clause 1, 2019) and in accordance with social responsibility, and should take 'appropriate action' to protect children or vulnerable adults from harm (UKCP clause 35, 2019).

Within a safeguarding context, assessment of risk and likelihood of injury tend to be considered on an individual level – the prospect of a client causing severe harm to themselves or another person. However, there are some actions that may be harmful on an individual level but nourishing on a collective one, and others that are collectively injurious and yet personally enjoyable – for example, collective action in demonstrations against climate change. Assessing risk in this context means exploring the client's thoughts, feelings and actions with sensitivity and openness, alert to the nuance between intention to act and fantasy.

Luepker (2022) indicates that our intention to safeguard against risk needs to also account for the risk to some clients of being exposed to potentially damaging social systems. Clients from marginalised groups, for example, are at greater risk of being systemically discriminated against, and our choices around the content, format and storage of client records should bear this in mind. This includes thinking carefully before storing notes via cloud technology, for example, and avoiding speculation in anything that could be subject to a subpoena.

In a time of climate breakdown, we must consider who – or what – we are intending to keep safe, and indeed what safety itself means. For many, the frameworks of our societies are likely to already be unsafe, whereas for others relative privilege means current systems are a safety net, although likely to be causing unconscious injury through a 'culture of uncare' (Weintrobe, 2020).

## Arrests and the Criminalisation of Protest

Attending a large environmental protest in London in 2023, a colleague confessed that even being present within the crowds felt like taking a risk. They said: 'I couldn't face arrest, like some activists; a criminal record would change my life, but it would also change so much for my clients'. We spoke about the UK's Police, Crime, Sentencing and Courts Act (2022) and the Public Order Act (2023), both of which criminalise protest-related actions, such as locking on and obstructing infrastructure, as well as more benign forms of public dissent, such as making noise during a protest and being in possession of materials intended for protest. My colleague continued: 'It feels right to speak up, but it is hard when the consequences extend beyond myself'.

Practitioners are encouraged to aspire to moral attributes, including being courageous enough to act in spite of risk and uncertainty (British Association for Counselling and Psychotherapy (BACP) clause 12, 2018). Many psychological professionals have joined protest marches and spoken publicly about issues of systemic inequality and climate breakdown. Some take part in non-violent direct action, such as joining slow marches or cracking the windows of fossil fuel-funding banks. Others participate in activism through offering free therapy sessions, or publishing materials beyond the gatekeeping of academia, in moves towards an anti-capitalist gift economy (George, 2022).

Many of our ethical frameworks explicitly state that practitioners who incur criminal charges must disclose this to their regulatory body and that criminal convictions may result in de-registration (BPC clause 4, 2011; BACP clause 47, 2018). This has had implications for practitioners who take an active stance in campaigning for climate action, particularly in countries which criminalise forms of protest, including the UK. Although thus far no such cases have led to practitioners being found unfit to practice, it remains a concern for many who are involved in civil disobedience and campaigns that contravene the government's direction.

Whilst the consequences of criminal charges in the UK do carry more weight with the introduction of recent legislation, activism can take many forms. What is urgently needed in order to deal with the changing landscape of the work is the need to revisit what it means to practise ethically. This applies just as much to us as individual practitioners as it does to our training providers and regulating bodies.

## Fitness to Practise

A trainee counsellor shared her concerns about continuing on her course as her awareness about ecological breakdown increased to a point where she felt she could no longer fail to act on it. She explained: 'I am worried that I'll end up facing a disciplinary hearing or a fitness to practise review because of speaking out about systemic problems and the way the accrediting bodies largely replicate the same structures'. We acknowledged that some psychological practitioners have had fitness to practise reviews, and that there have been reports of medical professionals risking being 'struck off' their respective registers, as well as medical students being asked to leave their training programmes.

The effects of ecological destruction – including the destruction of our own bodies through ingesting pollutants, exposure to heat stress, psychological overwhelm, among other factors – will form a necessary lens through which to assess our own fitness to practise in the context of climate breakdown.

To date, in all cases where a Fitness to Practice (FTP) hearing has been called due to arrest for non-violent protest, the HCPC has been unable to prove that FTP was impaired and it was not determined that such arrests were 'bringing the profession into disrepute' (BACP clause 48, 2018) but rather the opposite – that public regard was increased. It is important for therapists considering action to be aware of the risk and to request support and validation from their regulator (details cannot be provided for legal reasons).

Our fitness to practise in the context of climate breakdown is also related to our personal resilience. Enhancing our 'personal resilience' as the BACP (clause 3, 2018) exhorts us to do is likely not only to expand the accessibility of our practice, but also model a way of being in the world that is fluid, responsive and reparative. This means that the reality of climate breakdown cannot ethically be ignored or left out of the work; one cannot become more resilient to something unknown or unfaced (Richardson et al., 1990).

## Insurance

'I can no longer get insurance,' a client said, 'because of my criminal record'. They had been prosecuted for activities in relation to climate protests, which had subsequently impacted their ability to get home insurance. A colleague acknowledged the severity of this: 'It would be the same for a practitioner; we would find it very difficult, if not impossible, to get professional insurance with a criminal record'. We went on to discuss how other certification schemes, such as the Disclosure and Barring Service (DBS), in the UK would also be affected.

In private practice, loss of insurance is something a therapist would need to be explicit with clients about. It may be that some clients will choose not to work with a therapist who does not carry insurance, whilst others are likely to be more understanding, or not care. Either way, it would be ethical practice for the therapist to explain the implications of not holding insurance, such as the lack of indemnity in the event of complaints. Therapists may also experience difficulty renting a therapy room, and even lose eligibility for professional registration with the appropriate body (UKCP clause 39, 2019).

In a personal capacity, this is likely to also limit one's ability to hold vehicle insurance, buildings and contents insurance, life and health insurance. All of these have implications for managing professionally and personally within existing systems, such as invalidating a mortgage or limiting access to healthcare.

There are also emotional effects of professional or social exclusion, and sensitivity is required when working with clients experiencing loss of this kind. When therapists are experiencing this personally, we should engage in critical self-reflection and personal advocacy in order to ensure 'that our wellbeing is sufficient to sustain the quality of the work' (BACP clause 2, 2018).

Resourcing ourselves and our wellbeing in order to work with congruence and creativity within the context of climate breakdown is likely to require us to take an active role in connecting with our own needs. This may include practising whatever resources us as psychologically-informed individuals, connecting with our communities, and engaging with our own forms of Active Hope (Macy and Johnstone, 2012).

## Employment

'I support your ethics on a personal level,' an employer told me, 'but as an organisation, we could not support you professionally if you had a criminal record'. It was part of a conversation about my involvement in climate protests, the risk of arrest in light of changed legislation in the UK, and the terms of my employment contract. 'You will find that most organisations in this sector have similar policies,' this employer stated.

For practitioners in employment rather than private practice, contracts in the UK sometimes contain a clause specifically necessitating a 'clear' DBS certificate, or otherwise outline a prohibition on working with unspent criminal convictions. Employers may also have stipulations on their own insurances which prevent them from employing therapists with a criminal record.

Some employers may be uncomfortable with practitioners engaging in protest or activism, even when it does not lead to a criminal record. This could be argued to be a form of defensive collusion, where societal norms and assumptions have been internalised, shoring up pre-existing structures which inhibit the relationships and meaning-making processes (Bond, 2020). It may also stem from the organisation's failure to address anxieties around ecological destruction, or how the necessary changes to mitigate the worst effects of climate breakdown will also change the organisation itself. The social defences found in organisations are explored in depth elsewhere in this book (see Chapter 20).

## Discussion

Underpinning this chapter is the understanding that some ethical principles may conflict at times, and the emphasis is for practitioners to reflect on all ethical considerations before reaching a course of action. Guidelines state that 'practitioners may encounter circumstances in which it is impossible to reconcile all the applicable principles. This may require choosing which principles to prioritise' (BACP clause 7, 2018).

It is often the case that we respond to ethical quandaries through intuition or self-interest (Paxton, 2021), but therapists living in high-income countries exist within the same societal framework of neo-liberal capitalism as our clients. Thinking about how we engage ethically in an ecological crisis necessitates confronting these internalised systems.

A root cause of ecological breakdown is capitalism, a social structure that, it could be argued, alienates the best interests of the individual and the collective in favour of those of the corporation. As a result, the quality of connectedness within human relations has been eroded in many Global North societies; the individual is expected to sublimate need and desire in the name of an authoritarian Superego (Buzby, 2013). Working in this context means that taking appropriate action is likely to mean challenging capitalist socio-cultural norms.

Psychological practice can act as a handmaiden for the status quo, sometimes being actively or passively harmful. This can be seen in therapeutic expectations rooted in socio-historical customs rather than deliberative thought. For example, much therapy takes place one to one, in a quiet room where therapist and client sit opposite one another; there are social assumptions about what it means for a client to be 'functioning' and what is appropriate in terms of emotional display. Therapy can focus on enabling clients to take up paid employment or return to study as a marker of wellbeing; and there is often a presumption that a romantic or sexual relationship is a prerequisite for a healthy adult.

These are assumptions that some therapists take for granted, standards against which clients are measured in order to assess psychological wellbeing. It is this which has risked clients being pathologised for experiencing significant grief, anxiety or distress linked to their awareness of biodiversity loss, ecological destruction and climate breakdown (Hoggett, 2019). Responding to climate-related psychological effects in this way results in gaslighting the client and individualising what is a global issue.

Clients who express eco-anxiety and related feelings associated with biodiversity loss or climate breakdown often say that they have tried to explore these feelings with other professionals but have found their voices suppressed in favour of the therapist's theoretically-informed bias (Tait et al., 2022). Therapists themselves may be overwhelmed by their own climate-related feelings, and thus take up a defensive position, falling back on theory – and perhaps on ethical frameworks – to justify failing to engage with issues of climate justice in the therapy room.

Some therapists continue to interpret feelings about the external world solely as representations of early relational experiences. Such erasure of a client's felt experience stands in stark contrast with the ethical principle of non-maleficence (BACP clause 5, 2018) and the respect, integrity and dignity of the practitioner-client relationship (BPS clause 14, 2021). Study after study has shown that the relationship is the most important factor in the efficacy of therapy (Stargell, 2017).

In considering how to be a therapist during a time of climate and ecological breakdown, we must also consider our relationship with our own environment, including the model of therapy itself. Therapists exist within the structures and systems that both cause and are affected by environmental destruction. If we are to reflect on the tension inherent in such a position, and indeed to take up a more integrated way of being in the world (and in the therapy), we must grapple with those aspects of these systems which have been internalised. This may involve steps such as engaging with relevant CPD, perhaps returning to therapy ourselves, and connecting with a network of similarly climate aware practitioners.

## Conclusion

Paulo Coelho writes: 'The boat is safer anchored at the port; but that's not the aim of boats' (1992 [1987]).

By inviting in the messy and destructive forces at play in ecological destruction, we allow ourselves and our clients to work towards a state of integration, recognising that the dichotomy of safe therapy room and unsafe world is a false one. We navigate along a spectrum of safety and unsafety, a constantly changing dynamic with the possibility for creativity, destruction and all that lies between.

Therapy can seem safer when we are practising in a more certain world, and we can rely on our ethical frameworks to contain what can otherwise be a messy and unsafe process.

The existing ethical frameworks of our various professional bodies do allow for some interpretation and diversified culture of care (Weintrobe, 2020). But given the emergent and shifting nature of the multiple crises being faced, this chapter argues for a revised examination of these ethical frames. The world is changing rapidly, and therapy needs to change too.

## Note

1 All client details have been changed and amalgamated to preserve confidentiality.

## References

British Association for Counselling and Psychotherapy (2018) Ethical Framework for the Counselling Professions. Available at: https://www.bacp.co.uk/events-and-resources/ethics-and-standards/ethical-framework-for-the-counselling-professions/ [Accessed on 07/01/2023].

Barnett, J. (2011) Psychotherapist Self-disclosure: Ethical and Clinical Considerations. *Psychotherapy*, 48(4): 315–321.

Bond, P. (2020) Wounded Healer Therapists: A Legacy of Developmental Trauma. *European Journal for Qualitative Research in Psychotherapy*, 10: 68–81.

British Psychoanalytic Council (2011) Ethical Framework. Available at: www.bpc.org.uk/professionals/registrants/ethical-framework/ [Accessed on 02/02/2023].

British Psychological Society (2021) Code of Ethics and Conduct. Available at: www.bps.org.uk/guideline/code-ethics-and-conduct [Accessed on 07/01/2023].

Buzby, A. (2013) Wrong Life Lived Rightly: Sublimation, Identification, and the Restoration of Subjectivity. In: A. Buzby (ed.) *Subterranean Politics and Freud's Legacy*. New York: Palgrave Macmillan.

Coelho, P. (1992 [1987]) *The Pilgrimage*. San Francisco: Harper Collins.
Feltham, C., Winter, L. A, and Hanley, T. (2017) *The SAGE Handbook of Counselling and Psychotherapy*. London: SAGE Publications.
George, H. (2022) Taking Black Mental Health Matters into Their Own Hands. *Therapy Today*. Available at: www.bacp.co.uk/bacp-journals/therapy-today/2022/october-2022/the-big-issue/ [Accessed 30 June 2023].
Guterres, A. (2022) Transcript of Video Message to Launch Intergovernmental Panel on Climate Change (IPCC) report. Available at: https://press.un.org/en/2022/sgsm21228.doc.htm [Accessed 6 July 2023].
Gutheil, T. (2020) Ethical Aspects of Self Disclosure in Psychotherapy. *Psychiatric Times*, 27(5).
Hargaden, H. (2015) An Analysis of the Use of the Therapist's 'Vulnerable Self' and the Significance of the Cross-fertilization of Humanistic and Jungian Theory in the Development of the Relational Approach. *Self & Society: An International Journal for Humanistic Psychology*, 43(3): 208–212.
Health and Care Professions Council (2016) Standards of Conduct, Performance and Ethics. Available at: www.hcpc-uk.org/globalassets/resources/standards/standards-of-conduct-performance-and-ethics.pdf [Accessed 7 July 2023].
Hickman, C., Marks, E., Pihkala, P. et al. (2021) Young People's Voices on Climate Anxiety, Government Betrayal and Moral Injury: A Global Phenomenon. *The Lancet Planetary Health*, 5(12): 863–873.
Hoggett, P. (ed.) (2019) *Climate Psychology: on Indifference to Disaster*. London: Palgrave Macmillan.
Lewin, S. (2020) Schizoid Shame: The Idealization of Absence. *Contemporary Psychoanalysis*, 56(4): 534–561.
Luepker, E. (2022) *Record Keeping in Psychotherapy and Counselling: Ethics, Practice and Supervision. Oxon*: Routledge.
Macy, J. and Johnstone, C. (2012) *Active Hope: How to Face the Mess We're in Without Going Crazy*. San Francisco: New World Library.
Mighetto, I. (2022) 'Touching the Depths of Suffering with Others': The Nature of Witnessing with Survivors of Torture. *European Journal of Psychotherapy and Counselling*, 24(3): 315–338.
Orange, D. (2019) *Psychoanalysis, History and Radical Ethics*. Oxon: Routledge.
Paxton, R. (2021) Thinking About Thinking About Ethics. Available at: www.bps.org.uk/blog/thinking-about-thinking-about-ethics [Accessed 7 July 2023].
Police, Crime, Sentencing and Courts Act (2022). Available at: www.legislation.gov.uk/ukpga/2022/32/contents [Accessed 10 July 2023].
Public Order Act (2023). Available at: www.legislation.gov.uk/ukpga/2023/15/contents [Accessed 11 July 2023].
Richardson, G., Neiger, B. L., Jensen, S. et al. (1990) The Resiliency Model. *Health Education*, 21(6): 33–39.
Stargell, N. (2017) Therapeutic Relationship and Outcome Effectiveness: Implications for Counselor Educators. *Journal of Counselor Preparation and Supervision*, 9(2).
Tait, A., O'Gorman, J., Nestor, R. and Anderson, J. (2022) Understanding and Responding to the Climate and Ecological Emergency: The Role of the Psychotherapist. *British Journal of Psychotherapy*, 38(4): 770–779.

United Kingdom Council for Psychotherapy (2019) Standards, Guidance and Policies. Available at: www.psychotherapy.org.uk/ukcp-members/standards-guidance-and-policies/ [Accessed 13 April 2023].

Weingarten, K. (2003) *Common Shock: Witnessing Violence Every Day.* New York: New American Library.

Weintrobe, S. (2020) Moral Injury, the Culture of Uncare and the climate bubble. *Journal of Social Work Practice*, 34(4): 351–362.

Voice 4

# Timothy Morton – talking about climate agony, trauma and activism

Timothy Morton is Professor in English at Rice University. A member of the object-oriented philosophy movement, Morton's work explores the intersection of object-oriented thought and ecological studies, he is well known for development of the term hyperobject as it relates to climate change. Here he is talking with Caroline Hickman.

**Timothy:** Really, we should be talking about climate agony rather than climate anxiety, we are facing eco-trauma collectively now. We are all full of grief. We have gaslit ourselves into a dissociative state in relationship with other beings. We have been running from ourselves for over 50 years in denial about the way we are affecting the world and ourselves.

**Caroline:** Say a bit more about how you think we have gaslit ourselves.

**Timothy:** We are dealing with the rise in fascism, racism, homophobia, ecocide, alongside and as part of the climate and biodiversity crisis but collectively we refuse to face the full horror of all this. We are feeling huge shame and guilt about what we've done, and of course we need to defend against this. We are just now starting to wake up to this trauma. That's what we see as climate anxiety, but it is not just anxiety is it? We are in a fractured relationship with the non-human world, and with each other. The tensions are growing – we are all feeling this tension and it's about to pop. We should be afraid of this. What is Soul crying out for here?

**Caroline:** Has this gaslighting meant that we have failed to see Soul in this? You are talking about 'othering' on so many different levels, between black and white, between older and younger generations, between people and non-human others.

**Timothy:** That's where the Soul of this comes from - we can't see everything all at once, and that opens us up to this grief. This hyperobject. People tell me to 'get over it' – but I ask, 'why would you want to? Get over what? Get over what it means to live in this world today?'

**Caroline:** You are really moved emotionally by this, you look, and sound distressed. Can you say more about your climate grief?

**Timothy:** People like me are realising that we made it all worse. Every Greek plot is based on trying to escape my/our fate, but we have now created

DOI: 10.4324/9781003436096-9

this tragedy, it cannot be avoided. Because why? We know what to do, but we don't know why we have created this tragedy, so we are not doing what we should do. We need to examine the forms of activism that can express this for us now. Activists are finding ways to show us what is going on deep in our psyche.

When activists threw the soup at the paintings it was an expression of rage and despair on many different levels. It was both brilliant and nonsense at the same time. Nonsense because it confronts us with the fact that we value a painting higher than the life of our children. How mad is that? Activists need to see through this madness and find ways to act that expose this craziness. That soup action was brilliant and punk. There needs to be a Sex Pistols approach to environmentalism now. Less 'suits talking in private rooms' doing deals to seal the fate of the world's children behind closed doors, more soup.

Activists need to continue taking action in ways that have both meaning and nonsense and confront us with no meaning at the same time, they need to speak to this impossible binary. They also need to face the disgust of the public in order to have an impact. The emotions we are all feeling (or should be feeling) need activists to enact them in the world for us so we can relate to them, laugh, cry, applaud, despair. All of the feelings. The activists who threw the soup understood 'yes, we are in hell, but we are ecologically committed, and need to show people this'.

Psychologically this may not be a contradiction, activists are not contradicting all the feelings we are talking about here. They have the imagination to feel them all with these actions. They are striving to find the joy in the pain with these actions.

**Caroline:** Can we talk more about the relationship between activism and emotion?

**Timothy:** Environmental activism is a way of trying to be truly 'grown up', even though it is often framed as infantile and annoying. The main argument against it is often that it stops people from living their 'everyday lives'. These activists who are creating public spectacles of art with paint and powder are owning their own feelings about the crisis and showing the rest of us how to feel about the crisis. Not just talking about it. They are working with the unconscious mind to communicate to everyone else, even if this is not always consciously known. To find a way to 'be with' the climate crisis and wake people up we need to speak somehow to people's unconscious minds, we need to get everyone to see that all the cool kids care about climate change.

This is part of the dark ecology.

People (all of us) want to know what to do with all the awful feelings we are having about the state of the world today. We need to find ways through both psychology and activism to talk to people who are already upset, but in denial or pushing the feelings away, or they don't even know they are feeling them, all the people who don't know they're upset.

We are all wounded and broken but or by? trying to be ok in the world today; some people turn towards spending more, fighting and fucking more, some people turn towards caring more. We are all hurting and enslaving each other, let alone other animals. Maybe we are semi-conscious of this dynamic, we are aware that we are hurt, but activists show their hurt by trying to help others wake up, and take action to save others, while others use their hurt to wound others further.

How cool would it be if we all loved each other more?

**Caroline:** We are talking about the false binary of love and hate. Are you saying that activism holds the tension of these opposites?

**Timothy:** In order to address the climate and biodiversity crisis we also have to face the intersectionality that these crises are rooted in. We cannot have a classic white male saviour attitude and expect this to work now, when it has caused many of the problems we are facing. That would be absurd, wouldn't it? I have spent the last couple of years exploring my non-binary sexuality and can no longer separate out this from climate care or activism, or beauty or love or nature or art. Life is saying to me today that 'we live in a beautiful accident'. Embrace both. Always.

**Caroline:** So, what role do therapists have in this world, would you say?

**Timothy:** Climate aware – facing or knowing – therapists have a foundational role in so many ways. Free association is one of the only ways that we have to help us understand what we are living through with the climate crisis. Free association is a neurological lubricant. It helps us cry and laugh and throw paint or soup. If we can imagine our way forward through therapy and activism, maybe we can find our way forwards more generally.

**Caroline:** Crying and laughing sounds like what Sally Weintrobe describes as us needing a 'dream and a nightmare' to find our way forwards.

**Timothy:** Let's ask, what is the aim of the therapy session? I know something as this wounded person that other people don't know. As a survivor of child sexual abuse, I have always known things that others don't, or won't know. If we have gaslit ourselves around the climate crisis, then we have done that and are doing that for a reason. This is unbearable, the same as child sexual abuse, and of course we want to look away.

**Caroline:** So, what do we need to do now to look towards this, to face the unbearable nature of this? That is a shocking analogy to equate inaction on the climate crisis with turning a blind eye to child sexual abuse. I can understand that, but I wonder how we could make that understandable to others who maybe see humanity as separate from the planet.

**Timothy:** We need kind, very strong people. Therapists and activists – and maybe they are the same person – and we need the therapeutic process to inform the political process. Maybe these are the same.

**Caroline:** You seem to be suggesting that climate aware therapy is a form of activism.

**Timothy:** We need a therapeutic approach to everything that we do in facing the climate crisis, through activism, art, protest, politics ... what would it look like to have a therapeutic approach to politics? Instead of saying 'don't mention the war', we need to find radical strength and punk-heartedness in recognising that we are made through trauma – the recognition and understanding of trauma can give us empathy and compassion. It can soothe the oscillation between the hatred for the natural world and the love that drives activism. It's all about love in the end. That's what drives people to put their lives on the line, to risk their lives, risk their bodies on slow marches, and show their woundedness to the world in their despair and rage and pain.

We are living in absurd times.

**Caroline:** You seem to be saying something like, for activism to have an effect it has to have an edge of the absurd – orange paint or soup – as a way to communicate the ritual and emotion being evoked in humanity through the climate crisis. You have also drawn connections between your own journey of healing and recovery and identity, to the journey that we are all on facing this changing world today. Is there a way for us to draw some of these strands together without having to explain a soulful healing process, without making things too neat. Soup and paint are wonderfully messy. But can we find some relational final message? Or maybe we shouldn't do this.

**Timothy:** I'm not sure we should say more, leave the end of the conversation messy.

# Section Two

# Systemic Understandings

Chapter 4

# How Wide is the Field? Psychotherapy, Capitalism and the More Than Human World[1]

*Steffi Bednarek*

## Introduction

In 2018, the UN Intergovernmental Panel on Climate Change published a report calling for urgent changes at a global scale in order to limit the catastrophic effects climate change will have on life as we know it (Watts, 2018).

We may ask what this has got to do with psychotherapy and the mental health professions and how we can, as a discipline, begin to determine our role in wider world issues. The systems theorist Fritjof Capra (1982) points out that the dysfunction of complex systems on the world stage is primarily a crisis of *perception.* He believes that our seemingly innocent collective everyday beliefs contribute to the stuckness of much larger, complex systems. Our assumptions often serve as the connective tissue that holds things in their rigid place.

The Western culture of late-stage capitalism erodes the basis on which our existence rests, whilst psychotherapy continues to focus on our individual and inner worlds. Despite many theories that emphasise the individual's reciprocal interdependence with the world, our profession seems relatively ill-prepared to include the state of the world in our practice. It is the business of psychotherapists to bring awareness to fixed, dysfunctional and repetitive patterns. This chapter focuses on the patterns of thinking of psychotherapy itself and asks if there are areas in which our profession unconsciously reinforces assumptions that risk costing us the Earth. Are there aspects of our theories that inadvertently contribute to larger problems?

The birth of psychotherapy itself coincides with the height of the industrial revolution and the British Empire. Here, I explore aspects of the profession's alignment with the capitalist paradigm in its emphasis on anthropocentrism, progress, privatisation, domestication and materialism. I argue that this lens has become so familiar that we can hardly see beyond it. My main focus is on the relationship our discipline has with what cultural ecologist David Abram (1997) calls the 'more-than-human world', meaning the living world that humans are made of and participate in.

DOI: 10.4324/9781003436096-11

My intention is not to deliver a definitive statement, but to widen the conversation. I believe that psychotherapy has got something to contribute to the bigger picture but will not deliver 'the answer'. No single discipline can do this in isolation.

The philosopher Zygmunt Bauman (2000) writes about our Western culture as being in a state of 'liquid modernity' – a state that is characterised by chaotic, ungovernable situations, where change in one area of the system has ramifications throughout in unpredictable ways (Bednarek, 2017). Complex problems cannot be solved in old, familiar ways; they require transdisciplinarity. We have to think outside of the confines of the known, look beyond the boundaries of our defined schools of thought and widen the perspective.

Fritjof Capra (Capra and Luisi, 2014) tells us that in order to encourage change within a system, we do not have to lose all that we know. He suggests that change emerges from something dynamic within the old system, and advises us to pay close attention to emergence on the fringes. We can do this with our attention on large or small complex systems.

I suggest that we urgently need to widen our theories, open up dialogue and create new visions whilst staying clear of the old patriarchal ways of generating these visions, that is, where a few – often powerful, white, Western, upper or middle-class, male individuals – are in charge of the narrative. The celebration of the lone heroic thinker who presents 'the answer' denies co-created wisdom and is in itself a symptom of a patriarchal worldview that is letting us down.

## Response-ability in the Therapy Room

As an undergraduate student, I was a research assistant for an award-winning community psychology project in Germany. A team of psychotherapists worked in a deprived suburb. Instead of focusing on the 'dysfunction' in the individual, the attention was firmly on the cultural and collective malaise that revealed itself through personal stories. Once the community story had emerged, the team stepped out of conventional psychotherapeutic roles, lobbied for the community, negotiated with the local government, convened community gatherings and supported spontaneously emergent initiatives within the community. The structure, health and living conditions for the entire community improved measurably, which in turn impacted individuals' mental health. This experience inspired me to study psychotherapy.

However, it was impossible to find psychotherapy training that did not focus entirely on the dyadic relationship. Sociopolitical issues were understood to be part of the field but were hardly ever addressed directly. Whilst I became increasingly focused on understanding inner processes, attachment patterns and the private needs of my clients, the state of the Earth deteriorated. The assumption that more psychotherapy will automatically lead to a better world requires continual interrogation (Hillman and Ventura, 1992).

I suggest that before we can even begin to explore where our response-ability in the therapy room and beyond could lie, we need to understand where we may be aligned with the cultural values of the patriarchal and capitalist paradigm at the heart of major problems.

## Mental Health – A Cultural Construct

In a report entitled Regenerative Capitalism, the Capital Institute stresses a link between the capitalist worldview and global challenges such as climate change and political instability (Confino, 2015). According to this report, we need a move away from capitalist values towards a new systems-based mindset.

Western notions of mental health are far from neutral. Psychotherapy itself has 'grown up' in a capitalist, patriarchal and colonial Western culture that has inevitably left its imprint on it. What we have come to expect as 'healthy' human behaviour in a capitalist society is intrinsically linked to the dominant norms and values that govern our everyday lives (Rosenthal, 2008).

In a capitalist society we value individualism over community, ownership over an idea of the commons and private property over stewardship. We invest in growth models and operate in a competitive climate that tells us that we are masters of our own destiny if we only try hard enough. The Industrial Revolution has torn people from their communal contexts and their connections to place. The machinery of capitalist growth needed people to leave their rural communities in exchange for wage labour in cities.

The capitalist system requires a certain type of disposition in the general public in order to function (Adams, 2016). We are all expected to serve the capitalist model, to be motivated to contribute to the economy and to gain satisfaction and identity through a reward system that is removed from primary human needs. These culturally promoted characteristics have repercussions on our notion of expected behaviour and therefore also on what we perceive as deviations from the norm. Psychotherapists are not outside of cultural socialisation. We therefore have to reflect on the values that we perpetuate (consciously or unconsciously) and investigate where our understanding of health may inadvertently contribute to values that alienate us from each other and our participation with the world. In the following, I outline aspects of psychotherapy that have an uncomfortably close correlation with the dominant capitalist paradigm.

## Anthropocentric Worldview

Western culture mostly views the world from a perspective that places our species on top of the pile and all else beneath. Most psychological theories suggest that we are shaped by human relationships alone whilst our embeddedness in the more-than-human world is largely ignored. Our therapeutic discourse focuses almost entirely on human-to-human relationships. We do not include the absence of relationship with the living world in our diagnostic thinking of developmental trauma, attachment patterns, personality adaptations and mental health problems. Equally, our notion of community and relationship usually stops at the threshold of our own species. It rarely includes the relationship to trees, rivers, mountains, salmon, bees or water flowing through our bodies. What if we have anthropocentrised our understanding of human development at the cost of our belonging in the world?

Rarely does our thinking about trauma and loss include the catastrophic loss of attachment to place, the loss of endangered species, the loss of living in a functioning community, the loss of meaningful community rituals – even though these losses are so deep that they change who we believe we are. The idea that we may share a collective trauma is mainly unexplored.

We have become so inflated with our sense of ourselves as a species that we cannot see our actual dependence on that which we are destroying. The split between inner and outer has become 'normal' and therefore does not feature in our therapeutic theories or assessments.

The psychotherapist Francis Weller (2015) suggests that many of us carry a deep, but silent grief for our diminished sense of community with a world that we see as alive. We grieve for 'what it is we expected and did not receive' (2015: 54). In the absence of connection with a reciprocal world, we seek what we long for in our parents, partners or friends, who are expected to provide unconditional love and belonging. Many therapy sessions focus on the shortcomings of this expectation, suggesting that this longing can and should be met. The Earth is no longer experienced as Gaia (mother). Instead we look to our actual mothers, or to psychotherapists as the new mothers, to provide the magnitude that may be beyond human beings to provide. We have become literal about our need to be mothered whilst we have killed off and desecrated the much bigger feminine principle in our culture. And what we do not relate to, we are free to use, manipulate and destroy.

What would it be like to include the quality of our relationship with the more-than-human world in our assessments and our notions of trauma and attachment patterns? What is our profession's contribution to transforming this deep intergenerational disconnection?

## Boundary Between 'Me' and 'Not Me'

The question where the 'me' is located and where the 'other' begins and ends is at the core of psychological thinking. Where we see the boundary in our community with things determines the way we relate to the world. Most commonly we located this 'me' within the boundary of our physical skin. Postmodernism has deconstructed this rather simplistic idea of a coherent linear Self. Most psychological theories acknowledge that what we call 'I' is influenced and shaped by its contact with the world. Gestalt psychotherapy's theory of Self says that the Self is experienced at the contact boundary with wherever my focus is at any particular time. I can, for example, be so dissociated that my body feels fragmented and 'other'. Or I can be so confluent with someone that it is hard for me to distinguish a sense of Self at all.

Alan Boldon (2008) describes the absurdity of the way we think about the environment as 'other' in the attempt to try to decide at which point an apple we eat stops being part of the environment and becomes part of me, or at which point a raindrop that finds its way into the water I drink and then my body, is the environment.

The above illustrates that some psychological theories already reach beyond the anthropocentric paradigm. This is often implicit, but not explicitly explored in practice, and likely to stay a concept applied only to human-to-human interaction until we develop ways to work with what the Buddhist master Thich Nhat Hanh (1998) calls 'interbeing' – the essential interconnectedness and interdependence that binds us ever more deeply into the thick of the world.

This would imply that changes in the external world may be as therapeutic as changes in what we perceive to be our internal landscape, or that working on a client's feelings is not more or less therapeutic than working on cleaning a local riverbank.

Hillman (1995) suggests that the most radical intervention in psychotherapy would be a theory that replaces the individual with the world and that sees treatment of the inner requiring attention to be placed on the outer. This would be a departure from anthropocentrism and a decisive move towards a polycentrist view of life. Bill Plotkin (2013) proposes that an 'ecocentric' life means that all other memberships, such as primary partnerships, family, social groups, neighbourhood, workplace, profession, ethnic or gender identity group, state or nation become secondary to our inherent participation in the greater web of life. Our belonging in this web, and the wellbeing and care for it, become the primary concern and command the greatest loyalty.

If we think about the field of psychotherapy as an ecology of relationships, then the focus may shift away from efforts to delineate firm boundaries between clearly defined professional disciplines and move instead towards a trans-contextual cross-fertilisation, exploring the implications of a wider understanding of Self. This shift towards an ecological perspective may be a necessary condition for the evolution of the field at this consequential moment in time.

### Privatisation and Ownership

In capitalist cultures, we are taught to find comfort and identity in what we possess. Everything has become a commodity ready to be used for our benefit. Land is property, real estate, capital, recreation ground or natural resource. Even ideas, feelings, dreams or what goes on in our psyches are treated like private property.

As psychotherapists we encourage this sense of ownership by asking clients to 'own' their feelings, thoughts or ideas and talking about them as residing clearly within the client. In doing so our language has a norm of acquisition that separates us from the field context in which a particular feeling emerged. Once we own an idea, we can then extract the maximum potential from it, as if we were eternally hungry for something.

*What do you take from our session today?*

*What can I take from this dream?*

Many therapy sessions focus on what clients think they want, assuming that these needs deserve to be met. But maybe life calls us to serve something larger than our own individual needs, where the focus does not always 'fall on me' but on serving something outside of me and appreciating my participation in a bigger whole?

Weller (2015) believes that large aspects of Western culture have forgotten to think as a village or a commons and do not know anymore how to relativise the Self in service of community. The hegemonic ideology has isolated people out of a sense of belonging to a greater, more meaningful entity than their individual existence. The socialisation into the individualistic mindset means that most people do not even have ways of imagining a different way of being. And yet, a communal bond is integral to our human nature. We are wired for it.

An alternative approach would be to put our own lives in a relative position of service, allowing ourselves to surrender to it, serve its needs and be curious about what it wants from us rather than the other way around.

From this perspective we would ask what the dream, crisis or relationship asks of the client.

*What does this situation or problem require of you in order for you to do justice to it?*

*How can you be of service to the challenge that presented itself to you in your dream?*

Can we find in us the willingness to be of service to the things in life that are bigger than our own concerns, our own lifespan or the lifespan of the people we love? The question we need to address as therapists is how best to facilitate this reciprocal engagement with the world.

## Individuality

When asked about the role of the more-than-human world in the shaping of humanity, the human biologist Paul Shepard said: 'The grief and sense of loss, that we often interpret as a failure in our personality, is actually a feeling of emptiness where a beautiful and strange otherness should have been encountered' (Shepard, 1994: 214). In Shepard's opinion we have lost the continuity of connection to this beautiful and strange otherness to domestication. What follows is an emptiness. We typically blame ourselves for this feeling of emptiness and psychotherapy often colludes with this. Shepard asks us to consider that this emptiness may not be a personal shortcoming, privately owned, but a healthy reminder of something essential that we have lost in our encounter with the more-than-human world.

In a personalised psychology, based on individualism and ownership, we ascribe our feelings of emptiness to a failure in our own personality. The problem becomes interior and we try to fix or eradicate that which is calling out to us from beyond the confines of our individual lives. What we are left with is a chronic feeling of emptiness, which we can get so used to that we hardly feel it anymore. This makes it hard to name this permanent sense of loss.

Weller (2015) believes that what we are longing for are primary satisfactions, which evolved over thousands of years such as: gathering around communal life, story, mythology, meaningful relationships, ritual, fire, slowly evolving local connections, sharing and preparing food and spending time in nature. For the most part we have abandoned these primary satisfactions and are now surrounding ourselves with what he calls 'secondary satisfactions', like individual power, rank, prestige, wealth, status, material goods and stimulants, These are things that no matter how much we get of them, we will always want more to temporarily fill this permeating sense of emptiness.

The individualistic perspective tells us that we shape our own lives and that it is within our grasp to be content, unique and accomplished if we only try hard enough. This heroic ideal separates us from community and leaves us wide open to a sense of individual failure. As therapists we risk reinforcing this, by over-attending to a client's self-interests.

What if our primary human need is not to attend meticulously to our emotional wounds or to eradicate any signs of so-called mental health 'conditions', but rather to live our flawed and imperfect lives in a participatory way and to embody our fallible existence in deep connection with all that we encounter in the world?

## Addiction to Progress, Growth and Self-improvement

Capitalism likes things rising – stock markets, bank accounts, profit margins, house prices – whilst it is fearful of depression in the economy or in individuals. Those of us who live within a capitalist system are encouraged to focus on permanent improvement. We learn to fix and rectify that which resists an upward movement in a relentless pursuit of progress, and risk turning everything into a problem to be overcome – even death itself.

Aspects of our fallible human experience such as collapse, decay, loss, regression, stillness or stagnation are often approached with a notion of repair. It is therefore not surprising that many clients come to psychotherapy to create a Self that is approved by the world. This agenda is often based on self-hatred and a wish to eradicate the parts in them that stand in the way of the idea of progress and perfection. But in our attempts to domesticate that which frightens us, we risk pathologising the aspects of life that refuse to move anywhere or lead us downwards. This is the problematic aspect in the notion of 'healing' as opposed to an aesthetic approach that finds beauty in broken places.

The cultural obsession with things rising is often mirrored in psychotherapy when we collude with the idea of perpetual self-improvement or overemphasise the experience of lack, proclaiming that there has not been enough parenting, unconditional love, attachment and so on. In the hunger thus created lies the risk that both therapist and client are continually looking to fill up the emptiness. From this place, we devour the world without ever being nourished.

Most mythologies tell us that the price for initiation and wisdom has to be paid in the currency of suffering. An educated heart comes through the gateway of rupture. Bearing a certain level of pain is the vehicle that allows us to cross threshold moments. How can we facilitate this process for our clients, when therapist and client are both steeped in a grief-phobic culture?

Our experiences of abandonment, loss, death and betrayal are part of life and what has bound us together over centuries. Myths all over the world question not whether our hearts will get broken, but what meaning we ascribe to a broken heart. Do we follow the culturally dominant path of hunting for personal happiness or do we educate our hearts, allowing them to be broken, so that the world can flow into us?

## Materialism

I remember a psychotherapy session many years ago in which I expressed deep grief over a desolate landscape that I had visited that day. I expressed disgust at what we are doing to ourselves and the land we live on. After an exploration of where I felt this in my body, my experience was explored as a projection on the world. This can be important to explore, but it is a much-trodden path, often at the expense of other responses. The phenomenological exploration of my experience as perception or a dialogic encounter with place is extremely rare.

Is what is out there dead matter or is the more-than-human world able to communicate and reciprocate in its own way? Is the fact that we do not hear anything when we contact the world proof that there is no other consciousness than human consciousness or a sign that we have forgotten how to listen to a different language? The existence of non-human subjectivity is what indigenous cultures have lived by for millennia, but which the colonising cultures have eradicated. The question of matter holding consciousness is at the cutting edge of the current scientific debate. In philosophy, the concept of *panpsychism* holds the view that consciousness is a universal feature of all things (Bruntrup and Jaskolla, 2017).

The paradigm shift that we are faced with may not be an either/or, but seems to ask for a way to explore the field beyond polarities. We may continue to view subjectivity as only residing in human nature or we may expand our view of the field and consider the possibility of a subjectivity in animals, plants, waterways, trees and rocks. We are still a long way away from this. Do we, as psychotherapists, have anything to say about this?

## Lack of a Mythological and Cosmological Dimension

Descartes made the world dead. Everything has become solid matter. Unlike our ancestors, most white Europeans no longer feel at home with the mystical, mythical or divine. Most Westerners can no longer take their sorrows to a bigger entity.

Cosmology and mythology traditionally place the human experience in a wider context, but with the loss of connection to our mythological and cosmological ground we have become self-referential. It becomes about our individual lives. This risks creating a culture of literalness that becomes blind to that which is not tangible and dismisses meaning that is outside of our cognitive realm of reason. There are a few exceptions in psychotherapy that break with this norm.

In Jungian psychology the idea of soul, archetypes and the collective unconscious transcend the merely human realm ascribing agency to forces and presences outside of human control. Hillman (1995) suggests that if we want to live soulful lives we have to look outside of ourselves and engage with the 'anima mundi', the soul of the world. The anima mundi is an entity in its own right that acts upon us and asks us to participate. The Jungian perspective puts forward a view of the world that transcends the material and individualised perspective of the Western mind. The psychotherapeutic notion of a dialogic 'I-Thou' relationship is based on the work of Martin Buber (1958), who sees the premise of existence as encounter (Buber, 1947/2002). Buber's work was based on religious consciousness. He argued that an I-Thou relationship connects us in some way with the eternal relation to God. His work has influenced many psychotherapeutic modalities. though it seems that the profession has largely taken Buber's thinking and left God out of the picture.

The lack of a coherent exploration of what lies in the relational space between the 'me' and the 'not me' seems to suggest that many therapists do not ask this question. We leave it to individuals to decide whether they put a third entity in between. As in the wider culture, it has become a matter of personal taste and opinion. The lack of a cosmological and transpersonal perspective opens the profession up to the risk of practising a wild mix-and-match of individualised preferences. In our insatiable hunger we often appropriate other cultures' cosmologies and risk becoming consumers; taking whatever fills our longing for now and discarding what we do not like. We try Buddhism, shamanism, yoga, Sufism and so on, in the knowledge that we do not have to commit to any of it. We feel entitled to decontextualise that which is sacred to others.

In the absence of a transpersonal perspective, how do we learn to approach the world with a sense of wonder and where do we find our moral compass?

## Conclusion

I have outlined how anthropocentrism as well as the capitalist values of individualism, materialism, privatisation, ownership, progress and growth are reflected in our notion of mental health and the practice of psychotherapy in general. I highlighted that psychotherapists risk reinforcing a culturally endemic 'I-It' relationship to the world and argued that aspects of psychotherapeutic theory already lend themselves to a wider notion of the field. In order to build on the strengths of these theories, we have to expand our practice and rise beyond the individualistic, anthropocentric and atomistic paradigm.

When it comes to transitioning out of the deep rupture we have torn between us and the world, there are no rules, no maps and few elders to look to. It is up to us to figure out how to step out of our comfort zones and act in service of something that is greater than us.

When alarm bells are repeatedly ignored, the only way to wake up is through crisis. Some suggest that we are at the beginning of a major paradigm shift – a time of transition between the world as we have known it and a new world that we cannot yet know. This crisis may also be a threshold moment for the evolution of psychotherapy and our notion of what it means to be human.

## Note

1 This chapter is based on an already published article *British Gestalt Journal* 2018, 27(2): 8–17. Permission given.

## References

Abram, D. (1997) *The Spell of the Sensuous: Perception and Language in a More-Than-Human World*. New York: Vintage.

Adams, T. (2016) *The Psychopath Factory: How Capitalism Organises Empathy*. London: Repeater Books.

Bauman, Z. (2000) *Liquid Modernity*. Cambridge: Polity Press.

Bauman, Z. (2007) *Consuming Life*. Cambridge: Polity Press.

Bednarek, S. (2017) Therapy in a Changing World. *Therapy Today*, 27(1): 6–7.

Boldon, A. (2008) Climate Change – An Aesthetic Crisis? Available at: https://static1.squarespace.com/static/5ac4e0be4cde7a5b167238f8/t/65e63fd20727bf52a6aef672/1709588437039/Climate+Change+and+Aesthetic+Crisis+-+Alan+Boldon.pdf [Accessed 8 June 2018].

Bruntrup, G. and Jaskolla, L. (2017) *Panpsychism: Contemporary Perspectives*. Oxford: Oxford University Press.

Buber, M. (1947/2002) *Between Man and Man*. London: Routledge.

Buber, M. (1958) *I and Thou*. New York: Scribner.

Capra, F. (1982) *The Turning Point: Science, Society and the Rising Culture*. New York: Bantam Books.

Capra, F. and Luisi, P. L. (2014) *The Systems View of Life: A Unifying Vision*. Cambridge: Cambridge University Press.

Confino, J. (2015) Beyond Capitalism and Socialism: Could a New Economic Approach Save the Planet? *The Guardian*, 21 April 2015. Available at: www.theguardian.com/sustainable-business/2015/apr/21/regenerative-economyholism-economy-climate-change-inequality [Accessed 10 June 2018].

Hillman, J. (1995) A Psyche the Size of the Earth: A Psychological Foreword. In: T. Roszak, M. Gomes and A. Kenner (eds), *Ecopsychology: Restoring the Earth, Healing the Mind*. Berkeley, CA: Sierra Club Books.

Hillman, J. and Ventura, M. (1992) *We've Had a Hundred Years of Psychotherapy and the World's Getting Worse*. San Francisco: HarperCollins.

Nhat Hanh, T. (1998) *Interbeing. Fourteen Guidelines for Engaged Buddhism*. Berkeley, CA: Parallax Press.

Plotkin, B. (2003) *Soulcraft: Crossing into the Mysteries of Nature and Psyche*. Novato, CA: New World Library.

Rosenthal, S. (2008) *Sick and Sicker: Essays on Class, Health and Health Care*. (See ch. 3: Mental Illness or Social Sickness?)
Kindle edition: www.remarxpub.com, Remarx Publishing [Accessed 8 June 2018].

Shepard, P. (1994) The Unreturning Arrow. In: J. White (ed.) *Talking on the Water: Conversations about Nature and Creativity*. San Francisco: Sierra Club Books.

Watts, J. (2018) *We Have 12 Years to Limit Climate Change Catastrophe, Warns UN. The Guardian, 8 October 2018*. Available at: www.theguardian.com/environment/2018/oct/08/global-warming-must-not-exceed-15c-warns-landmark-unreport [Accessed 8 October 2018].

Weller, F. (2015) *The Wild Edge of Sorrow: Rituals of Renewal and the Sacred Work of Grief*. Berkeley, CA: North Atlantic Books.

Chapter 5

# Climate Distress through the Lens of the Power Threat Meaning Framework

*Gareth Morgan*

## The Power Threat Meaning Framework (PTMF[1])

The PTMF (Johnstone et al., 2018), freely available via the British Psychological Society,[2] was developed to offer an alternative to diagnosis-led systems of classification. Whilst a critique of diagnosis is beyond the scope of this chapter, the idea that there are discrete 'mental disorders' remains unevidenced (see PTMF Ch 1; Ch 5) and the medicalisation of human suffering has been linked to neoliberal ideologies that have accelerated our planetary emergencies (Davies, 2022; Weintrobe, 2021). Conversely, there is a large body of evidence linking various forms of adversity to human suffering (PTMF Ch 4). The PTMF was developed from this literature with the aim of supporting people to develop non-pathologising narratives, enabling them to recognise their suffering as understandable in the context of lived experience and toxic social structures.

There are five PTMF elements, outlined in Box 5.1. Although presented separately, the domains overlap and it is not suggested they should be worked through in a linear order: narrative development involves reflection over time, moving back and forth between the interrelated elements dialogically.

**Box 5.1 The five elements of the PTMF**

1 What has happened to you? (How is *power* operating in your life?)
2 How did this affect you? (What kinds of *threats* did this pose?)
3 What sense did you make of it? (What is the *meaning* of these situations and experiences for you?)
4 What did you do to survive? (What kind of *threat responses* were/are you using?)
5 What are your strengths? (What access to *power resources* do you have?)

DOI: 10.4324/9781003436096-12

## Power

An analysis of power should be central to any approach that aims to contextualise emotional suffering. Whilst 'power' is a contested concept, the PTMF offers tentative definitions: 'The means of obtaining security and advantages for yourself or others' and 'being able to influence your environment to meet your own needs and interests' (Boyle and Johnstone, 2020: 41). Table 5.2 illustrates one way of grouping power with examples for people concerned about climate breakdown within the UK. Ideological power is of particular importance for making sense of meanings (Boyle, 2022).

## Threat

When power operates negatively, it poses threats to our ability to meet core needs such as needs for safety, physical sustenance, to feel loved, to have proximity to those we care about and to feel valued within our communities and those important to us. Direct impacts of climate breakdown, such as extreme weather events, war, famine and displacement, pose extreme threats to basic human needs. Awareness of the real risks of social collapse and possible human extinction also present existential threats to people in the privileged position of not yet enduring acute impacts (Budziszewska and Jonsson, 2021). Recognition of the multiple existential threats faced can challenge worldviews most people in the UK have been socialised to accept: That humanity has mastery over nature; that (western) civilisation will continue to thrive; that powerful people and institutions act to protect our best interests (Gillespie, 2020).

People can encounter relational threats such as alienation if loved ones do not share concerns or ridicule their actions (Hoggett and Randall, 2018). Legal and financial penalties present further threats to activists' abilities to meet their own and loved ones' physical needs, and can result in dilemmas about whether to participate in different forms of activism, whilst threats of violence and disproportionate responses from police and private security firms can be heightened for those experiencing racism (Joseph-Salisbury, Connelly and Wangari-Jones, 2021).

Many of the questions listed in Table 5.1 may be useful for supporting the identification of threats because negative operations of power inevitably invoke threats. Follow-up questions relating to the Threat domains listed in Box 5.2 might support awareness of the effects of negative operations of power for a person.

## Meaning

Although humans are active in ascribing meaning to their experiences, the Framework draws on the works of Shotter to argue that the meanings

*Table 5.1* Power and climate breakdown

| *Form of power* | *Example links with climate breakdown* | *Example questions to support narrative development* |
|---|---|---|
| **Legal power**: the use of the law to support or inhibit needs/justice | Racial profiling; stop and search powers; legal clauses preventing activists speaking about the climate crisis when giving evidence; laws preventing/limiting protests; laws limiting refugees' rights; laws supporting the protection/destruction of the more-than-human world | How have laws/changes in laws impacted upon what actions/activism you've engaged in? In what way has the legal system enabled the harms to the planet that you are trying to prevent? |
| **Coercive power**: power by aggression or threat of violence | Arrest, assault, or political action against climate activists | How have the police/members of the public responded to you during actions? Have you experienced threats/violence from anywhere else? |
| **Embodied power**: *interactions between* a person's body and society (e.g., intellectual, social or physical abilities valued in a specific cultural context) | Limited mobility inhibits adaptation. Physical health/age moderates impact of extreme heat, pollution, impact of diseases etc. People with black or brown skin are at heightened risk of police violence/harsher sentences | If society was not structured in such an ableist/ageist/racist (etc.) way, what opportunities would this open up to you with regards actions/activism you might engage in?<br>In all cases, it is important to frame oppression as products of historically-rooted overt and institutional discrimination, not problems 'within' individuals/marginalised groups. |
| **Interpersonal power**: the power to get one's relational needs met; to support/deny others' needs | Whether loved ones validate or dismiss climate concerns; the degree to which activist groups offer support and understanding or have a culture that blames/invokes guilt | How have important others in your life reacted when you've shared your concerns about the planet? When you've spoken of your activism? Or encouraged them to take action?<br>Do worries about loved ones impact upon what actions/activism you do/don't engage with? Are there things that others in your activism group do that support you or make you feel worse? |

| | | |
|---|---|---|
| **Social capital**: power associated with a person's standing within a given community | Whether a person's concerns are heard or validated may relate to their standing, which will vary with factors such as age, education, profession, 'race', 'class', etc. | Do you think others respond differently because of your 'race'/'disability'/'class' etc.? If you didn't have a mental health diagnosis? How might others respond differently to you if you were a scientist/professional? If you were an adult/teenager? What actions would be available to you if you had a higher standing in society/your profession/your activist community? |
| **Economic power**: the ability to use economic and material resources | Money to: afford sustainable products; leave a job that goes against a person's values; participate in activism; donate to environmental causes; rebuild following extreme weather | What would you be doing with regards your activism/lifestyle if money wasn't a restriction? If you had enough money to look after your family for several years? |
| **Ecological power**: the degree to which local ecosystems support our needs; power to nurture local ecosystems | Degree to which local ecosystems provide food, access to green space, protection from flooding/wildfires/droughts/heatwaves | What changes have you noticed regarding climate and biodiversity loss? Are places/species important to you under threat? How much power do you have to protect/nurture these ecosystems? Have you witnessed the destruction of nature? |
| **Ideological power**: the control of meaning through dominant messages from society, media, social sciences, etc. | Messages that: deny/minimise threats posed by planetary emergencies; place burden for change on individuals, minimise/over-emphasise the effectiveness of particular actions/activism; position climate action as in opposition to improving 'quality of life' (Lamb et al., 2020).<br>Messages that position climate activists as 'dangerous radicals' or support their dismissal through stereotyping them as 'hippies'/'entitled'/'scroungers' etc.; position humans as separate (and | Ideological power can be difficult to identify as it operates to serve existing power structures and inequalities by making us think 'this is the way the world is'. Questions can be asked about awareness of the different messages in the previous column. Reviewing newspaper articles/cartoons/social media posts can support identification of ideological power in relation to relevant topics, e.g., portrayal of climate activists. Consideration can be given to other operations of ideological power relevant to a person's identity (see the *Identities* provisional-pattern in the PTMF) |

*(Continued)*

*Table 5.1* (Continued)

| *Form of power* | *Example links with climate breakdown* | *Example questions to support narrative development* |
| --- | --- | --- |
| | superior) to the more-than-human world; position peoples from Majority World nations as 'other'/'lesser'; privilege economic growth over planet and equality.<br>Messages about (un)healthy emotional responses to climate breakdown: that it is unequivocally harmful to speak to young people on this topic.<br>Messages about how we are 'meant' to feel/behave in wider society are also important: the pursuit of happiness through wealth, status and career development | |

Adapted from Johnstone et al., 2018; Morgan et al., 2022.

**Box 5.2 Example threats**

**Relationships** with relatives/friends/colleagues/other activists; including standing within social groups, amount of contact with others, threats of ridicule/marginalisation.

Abilities to regulate **overwhelming emotions** such as anger, grief, despair, hopelessness, guilt

**Economic:** loss of income associated with fines/loss of work

**Existential:** threats to sense of ongoing civilisation; to spirituality/faith; sense of purpose; trust that powerful others look after our interests

**Bodily:** violence towards activists; bodily impacts of chronic stress

**Values:** recognition that lifestyle has supported harm; threats to valuing legal/political systems

**Identity:** threats to what it means to be human, a UK citizen, someone of a particular profession/occupation, a parent/grandparent/someone who wanted children. Threats to identity associated with specific forms of activism; identifying/being identified as someone 'suffering' from 'climate anxiety'

available to people are constrained by language, culture and operations of ideological power (PTMF Ch 3). Rejecting the recent western-psychological trend that separates cognition from feelings, the Framework recognises that some meanings (e.g., shame, hopelessness) inevitably evoke thoughts, emotional responses, bodily sensations, memories and images.

When a person is able to stay connected to the multiple existential threats associated with climate breakdown, they may experience a range of meanings such as despair, fear, grief, meaninglessness, powerlessness and a sense of being betrayed by powerful others (Hickman et al., 2021; Wray, 2022). Some may wrestle with meanings of guilt or shame when recognising harmful impacts of lifestyles, or for not having recognised the urgency of our situation earlier (Pihkala, 2022; Weintrobe, 2022).

Whilst some may experience meaninglessness (Budziszewska and Jonsson, 2021; Woodbury, 2019), participation in activism can result in meanings related to newfound purpose and a sense of connectedness to others and the more-than-human-world (Macy and Johnstone, 2022). However, some may continue to experience guilt about 'not doing enough', or powerlessness/hopelessness when recognising the scale of the threats faced (Diffey et al., 2022; Sanson and Bellemo, 2021). If activists are described as 'suffering' from 'climate anxiety' or are diagnosed with a 'mental health condition' then there can be further risks that they internalise messages about what it means to be 'mentally ill'. In doing so, meanings such as defectiveness, shame or alienation might be experienced (PTMF: 221–222).

Some people find it easier than others to identify meanings; a list may assist. As with all aspects of the PTMF, a 'menu of meanings' is not exhaustive, and people can experience a range of contradictory meanings.

## Threat Responses

This element relates to the ways people respond to either avoid or mitigate threats. Example threat responses for people distressed by climate breakdown in the UK are listed in Table 5.2. The same threat response can serve different functions for different people or for the same person at different times. For example, *fatalism* – believing that things are hopeless – could be a response to protect against overwhelming guilt, or to protect against feeling powerless in the face of the magnitude of the climate emergency (Hickman, 2020). Similarly, participation in *activism* can serve different functions, such as protecting against emotional overwhelm, helping a person to feel in control, or providing a sense of purpose and protecting against anticipated dangers. Activism could also be driven by a desire to protect against overwhelming guilt or a sense of personal responsibility (Hoggett and Randall, 2018), or to feel valued within activist communities. The Framework invites a nuanced consideration of a range of threat responses that serve a variety of different functions for a person.

The threat responses a person adopts will vary depending on the meaning attributed to threats in their personal contexts (e.g., acceptability of different threat responses varies with culture), as well as the power resources available to them. For example, the ability to take part in differing forms of activism will be influenced by whether a person can afford to take time off work, travel to certain locations or risk the consequences of fines. Conversely, participation in civil disobedience may be experienced as one of the few available

*Table 5.2* Example threat responses

| *Example functions* | *Example threat responses* |
|---|---|
| Regulating overwhelming feelings | Dissociation; emotional numbing; distraction; giving up/withdrawing, fatalism; wishful thinking; denial/minimisation; intellectualising; activism; self-harm |
| Maintaining a sense of control/agency | Denial/minimisation; fatalism; compartmentalising; making lifestyle changes; activism; growing food |
| Protection from dangers | Fight/flight/freeze responses; hypervigilance (e.g., seeking news about the climate crisis); ruminating; growing food; pleading with others to take the crises seriously; activism |
| Striving for acceptance within valued social groups | Avoiding discussing climate change/activism; participating in forms of activism to maintain position/feel valued within activist communities; striving to do well in career/education |

threat responses for people who feel they have no other means to get powerful others to recognise the urgency of the situation and act accordingly.

Additional threat responses relating to climate breakdown are listed in Morgan et al. (2022), whilst the PTMF offers other common threat responses (PTMF: 212–213). The lists are non-exhaustive and there are potentially an infinite number of threat responses people will utilise. However, the lists might offer a useful starting place for supporting people to think about the different strategies they employ for different functions.

## Strengths

The final element of the PTMF focusses on a person's strengths. These can include the power resources available to a person, as well as values important to them. Meanings such as despair or rage might support the identification of such values, as these feelings can speak to what a person holds precious. Given the nature of threats posed by climate breakdown, there could be value in supporting a person to broaden consideration of strengths beyond the individual level to include attention to the resourcefulness of their communities. Recognition of the human capacity for extreme compassion and recognition that many others are taking action could support people to remain engaged with climate activism without feeling overwhelmed (see 'grounding in gratitude' in Chapter 17 of this book).

## Narrative Development and Utility

After exploring how the different elements of the PTMF relate to a person's experience, they might choose to integrate content to develop a narrative. The PTMF privileges narrative development over categorisation, as a narrative enables a person to explain their experiences and distress in their own language. The Framework can also be utilised to frame responses of communities, as illustrated in Barnwell, Stroud and Watson's (2020) case study concerning responses to the climate and ecological emergencies among a South African mining community.

Whilst the application of the Framework to people distressed by climate breakdown is currently being explored, accounts from survivors of mental health systems have highlighted how the process of narrative development can support people to make connections between threats, operations of power and personal experience in a way that reduces self-blame (Ball, Morgan and Haarmans, 2023). It is hoped the Framework might have similar benefits for those concerned about the multiple existential threats we face, and will also be of use in supporting services to take non-pathologising approaches to making sense of the distress and actions of those seeking support.

Other potential benefits of PTMF narrative development for an individual include:

- Reducing self-blame/overwhelming guilt for having lived lifestyles that are harmful to the planet; recognition of the various ways power has concealed the urgency of the climate and ecological emergencies.
- Reducing guilt for using denial/distraction; supporting recognition that individuals are not responsible; that avoidant threat responses serve functions that prevent 'burnout'.
- Making sense of distress in relation to climate breakdown as a strength; an ability to connect with the urgency of our time (Sangervo, Jylha and Pihkala, 2022).
- Legitimising non-Western and Indigenous wisdom about our interconnectedness with each other and the more-than-human world.
- Supporting reflection on the helpfulness of different threat responses – for the person, and for the planet.
- Recognising that threat responses are constrained by operations of power; and that intersectional oppression impacts upon how we respond.
- Supporting a sense of connection where others' narratives resonate with one's own.
- Supporting awareness that important others in a person's life who might appear unconcerned will also be responding with threat responses (see, the climate crisis as a collective trauma, Woodbury, 2019).

The following uses might be afforded for mental health services or activist groups concerned about members' wellbeing:

- Supporting awareness of the intersection of climate and ecological emergencies with other social justice issues (Barnwell and Wood, 2022; Ogunbode, 2022; Sultana, 2022).
- Recognising that neither distress nor participation in activism should be pathologised/problematised; it is more helpful to explore meanings, threat responses, etc. *collaboratively*; what is more or less helpful for the person/community/planet?
- Supporting professionals to reflect on their own threat responses to climate breakdown and what they might be bringing to a therapeutic relationship.
- Orienting services to the importance of amplifying the voice of those concerned about climate breakdown (Diffey et al., 2022; Li et al., 2022).
- Attending to the importance of community and bringing people together to support validation, empowerment, and movement towards the radical changes the IPCC (2022) has warned we need in all aspects of society.

## Conclusion

This chapter has offered some ideas about how the PTMF might be utilised with people distressed about climate breakdown. The focus has been on supporting people in a UK context. It is important to recognise that power, threat, meaning and threat responses will be impacted by culture and context. The UK to date has been spared the most acute impacts of climate breakdown. Whilst legal and coercive power has been ramped up against activists in the UK, hundreds of environmental defenders are murdered each year in other nations (Frontline Defenders, 2022).

The Framework should never be imposed upon someone (Boyle and Johnstone, 2020) and, as evidenced by the present book, is but one of many available models for making sense of climate related distress. It is hoped some individuals, activist communities and therapists might find the Framework useful for supporting self-other compassion, enabling people to take climate action without feeling overwhelmed, and supporting much-needed connection and community cohesion to navigate the unprecedented challenges we must face together (Mead et al., 2023).

## Notes

1 Content is condensed from a co-authored paper that considers a broader range of responses to climate breakdown (Morgan et al., 2022).
2 www.bps.org.uk/guideline/power-threat-meaning-framework-full-version

## References

Ball, M., Morgan, G. and Haarmans, M. (2023) The Power Threat Meaning Framework and 'Psychosis'. In: J. A. Diaz-Garrido, R. Zuniga, H. Laffite and E. Morris (eds) *Psychological Interventions for Psychosis: Towards a Paradigm Shift*. London: Springer.

Barnwell, G., Stroud, L. and Watson, M. (2020) Critical Reflections from South Africa: Using the Power Threat Meaning Framework to Place Climate-related Distress in its Socio-political Context. *Clinical Psychology Forum*, 332: 7–15.

Barnwell, G. and Wood, N. (2022) Climate Justice Is Central to Addressing the Climate Emergency's Psychological Consequences in the Global South: A Narrative Review. *South African Journal of Psychology*, *52*(4): 486–497.

Boyle, M. (2022) Power in the Power Threat Meaning Framework. *Journal of Constructivist Psychology*, 35(1): 27–40. 10.1080/10720537.2020.1773357.

Boyle, M. and Johnstone, L. (2020) *A Straight Talking Introduction to the Power Threat Meaning Framework*. Monmouth: PCCS Books.

Budziszewska, M. and Jonsson, S.E. (2021) From Climate Anxiety to Climate Action: An Existential Perspective on Climate Change Concerns Within Psychotherapy. *Journal of Humanistic Psychology*, 0(0). 10.1177/0022167821993243.

Diffey, J., Wright, S., Olachi Uchendu, J., et al. (2022) 'Not About Us Without Us' – The Feelings and Hopes of Climate-concerned Young People Around the World. *International Review of Psychiatry*, 34(5): 499–509. 10.1080/09540261.2022.2126297.

Davies, J. (2022) *Sedated: How Modern Capitalism Created Our Mental Health Crisis.* London: Atlantic Books.

Frontline Defenders (2022) *Global Analysis 2021.* Available at: www.frontlinedefenders.org/en/resource-publication/global-analysis-2021-0.

Gillespie, S. (2020) *Climate Crisis and Consciousness: Re-imagining Our World and Ourselves.* Oxon: Routledge.

Hickman, C. (2020) We Need to (Find a Way to) Talk About … Eco-anxiety. *Journal of Social Work Practice*, 34(4): 411–424.

Hickman, C., Marks, E., Pihkala, P. et al. (2021) Climate Anxiety in Children and Young People and Their Beliefs About Government Responses to Climate Change: A Global Survey. *The Lancet Planetary Health*, 5(12): 863–873. 10.1016/S2542-5196(21)00278-3.

Hoggett, P. and Randall, R. (2018) Engaging with Climate Change: Comparing the Cultures of Science and Activism. *Environ. Values*, 27: 223–243.

Intergovernmental Panel on Climate Change (IPCC) (2022) *Climate Change 2022: Mitigation of Climate Change.* Available at: /www.ipcc.ch/report/ar6/wg3/ [Accessed 28 June 2023].

Johnstone, L. and Boyle, M., with Cromby, J., Dillon, J., Harper, D., Kinderman, P., Longden, E., Pilgrim, D. and Read, J. (2018) *The Power Threat Meaning Framework: Towards the Identification of Patterns in Emotional Distress, Unusual Experiences and Troubled or Troubling Behaviour, as an Alternative to Functional Psychiatric Diagnosis.* British Psychological Society. Available at: www.bps.org.uk/member-networks/division-clinical-psychology/power-threat-meaning-framework [Accessed 9 May 2023].

Joseph-Salisbury, R., Connelly, L. and Wangari-Jones, P. (2021) 'The UK Is Not Innocent': Black Lives Matter, Policing and Abolition in the UK. *Equality, Diversity & Inclusion*, 40(1): 21–28. 10.1108/EDI-06-2020-0170.

Lamb, W., Mattioli, G., Levi, S., et al. (2020) Discourses of Climate Delay. *Global Sustainability*, 3: 17.

Li, C., Lawrance, E. L., Morgan, G., Brown, R., Greaves, N., Krzanowski, J., Samuel, S. and Guinto, R. R. (2022) The Role of Mental Health Professionals in the Climate Crisis: An Urgent Call to Action. *International Review of Psychiatry*, 34(5): 563–570. 10.1080/09540261.2022.2097005.

Macy, J. and Johnstone, C. (2022) *Active Hope: How to Face the Mess We're in with Unexpected Resilience and Creative Power* (second edition). San Francisco: New World Library.

Mead, J., Gibbs, K., Fisher, Z. and Kemp, A. H. (2023) What's Next for Wellbeing Science? Moving from the Anthropocene to the Symbiocene. *Frontiers in Psychology*, 14. 10.3389/fpsyg.2023.1087078.

Morgan, G., Barnwell, G., Johnstone, L., Shukla, K. and Mitchell, A. (2022) The Power Threat Meaning Framework and the Climate and Ecological Crises. *Psychology in Society*, 63: 83–109.

Ogunbode, C. A. (2022) Climate Justice Is Social Justice in the Global South. *Nat Hum Behav*, 6: 1443. 10.1038/s41562-022-01456-x.

Pihkala, P. (2022) Toward a Taxonomy of Climate Emotions. *Frontiers in Climate*, 3. 10.3389/fclim.2021.738154.

Sangervo, J., Jylha, K. M. and Pihkala, P. (2022) Climate Anxiety: Conceptual Considerations, and Connections with Climate Hope and Action. *Global Environmental Change*, 76. 10.1016/j.gloenvcha.2022.102569.

Sanson, A. and Bellemo, M. (2021) Children and Youth in the Climate Crisis. *BJPsych Bulletin*, 45(4): 205–209. 10.1192/bjb.2021.16.
Sultana, F. (2022) The Unbearable Heaviness of Climate Coloniality. *Political Geography*, 99. 10.1016/j.polgeo.2022.102638.
Weintrobe, S. (2021) *Psychological Roots of the Climate Crisis: Neoliberal Exceptionalism and the Culture of Uncare*. London: Bloomsbury.
Weintrobe, S. (2022) The New Bold Imagination Needed to Repair and Expand the Ecological Self. In: W. Holloway, P. Hoggett, C. Robertson and S. Weintobe (eds) *Climate Psychology: A Matter of Life and Death*. Bicester: Phoenix.
Woodbury, Z. (2019) Climate Trauma: Toward a New Taxonomy of Trauma. *Ecopsychology*, 11(1): 1–8.
Wray, B. (2022) *Generation Dread: Finding Purpose in an Age of Climate Crisis*. Canada: Knopf Canada.

Voice 5

# Disability and climate anxiety

*Helen Leonard-Williams*

My experiences of climate anxiety deeply intersect with my experiences of disability and ill health. Due to the ableism we face at every level of society, disabled people are all too often treated as disposable in crisis situations and are much more likely to live in poverty and be isolated from our communities. This makes us particularly vulnerable to climate change and less able to access support for our mental health, including the deep feelings of anxiety, fear, anger and grief surrounding the climate crisis many of us experience. This lack of support is often exacerbated if we are multiply marginalised.

For me, feelings of fear, anger, grief and anxiety about the climate crisis started at the same time I contracted myalgic encephalomyelitis (ME), a debilitating neurological disease, in 2017. Before then my plan was to help tackle the climate crisis through both individual action and future work; I felt like I was in control of my future and had a plan to help do my part to protect people and our planet. After getting ME, I became more afraid and angry that so many people in positions of power were failing us but felt unable to take action due to being ill and feeling cut off from the often very able-bodied and ableist mainstream climate movement. I became angry at and disillusioned with the climate movement as I didn't feel like it was including or fighting for marginalised disabled lives. It was an incredibly isolating and disempowering few years.

The systems supposedly supporting disabled people in this country have been described by the UN as a 'human catastrophe' and a deep violation of our rights, as Frances Ryan (2019) describes. For example, in 2020 I attempted to apply for disability benefits. It took nearlytwo years of advocacy and energy to be awarded them and was the most dehumanising thing I'd ever been through. My mental health suffered greatly. Every system supposed to help us survive (healthcare, government, education, work, social care etc.) is at best neglecting millions of disabled people across the country, and there is rational fear and anxiety that climate change will exacerbate this. We fear that we, disabled people who need more – or different – support than nondisabled people to thrive, will be left behind; treated as acceptable losses. These thoughts are terrifying but unfortunately not unfounded – we're

DOI: 10.4324/9781003436096-13

already seeing this happen as clinically vulnerable people are forced to isolate from society and go without social care, medical treatment, other services, and social gatherings to avoid the ongoing pandemic; there are now no mandated protections against.

Climate change will also exacerbate many disabled people's medical conditions, for example, we may be less able to withstand extreme heat or cold and less able to protect ourselves against them due to poverty. Personally, the possibility of my medical conditions becoming more severe due to climate change coupled with the knowledge that there isn't currently any support in place for people with more severe ME is hugely frightening and a big factor that contributes to my climate anxiety.

The amount of energy it takes to simply survive day-to-day as a disabled person can mean we have no energy left to access community, increasing feelings of loneliness, anxiety and fear. It's utterly exhausting, physically and emotionally. Joining the UK Youth Climate Coalition in 2021 gave me a vital community where I can organise most weeks with other young people passionate about climate justice in an accessible environment to me, and my climate anxiety has improved. However, so many disabled people still cannot access such vital spaces, and this must be recognised when supporting us.

When accessing counselling, I still find that practitioners don't fully understand how intertwined climate anxiety is with my experiences as a disabled, chronically ill person. This is a huge failure on their part. It worsens isolation and if anything, interactions with practitioners can increase climate anxiety rather than help with it. Practitioners wanting to support disabled people with our climate anxiety must learn about or have training on the specific ways we are affected by climate change, ableism, community isolation and the neglect we too often experience from the DWP, healthcare providers – including psychologists – social care, workplaces, family, friends and more. Practitioners must work to combat the unacceptable lack of access to mental health care disabled people experience and the ways their practices are inaccessible to marginalised disabled people most affected by climate change, for example, due to expense, or a lack of online or messaging options, COVID precautions, or wheelchair access. Actively dismantling their own ableism and ways in which ableism shows up in their practice must be essential, continual work.

Helen Leonard-Williams (they/them) United Kingdom Youth Climate Council (UKYCC)

## Reference

Ryan, F. (2019) *Crippled: Austerity and the Demonization of Disabled People*. London: Verso.

Chapter 6

# Deep Democracy: World Out There – World in Here

*Iona Fredenburgh and Sue Milner*

## Introduction

Our focus in this chapter is the entanglement of the world 'out there' and the world 'in here'. They are not separate. As therapists we are not only in a one-to-one relationship with our client, but with the world itself, through everything our clients are and do, whether they are aware of that or not. Likewise, the world is implicit in our own experiences and perceptions as therapists: we are not neutral, and we need the breadth of that awareness in our practice, with its different facets, while also addressing the specific individual process of the client. Similarly, for those of us engaging with social activism, for instance around climate justice: we need to include an understanding of the link between outer systemic processes and our own inner world, and the inner worlds of others.

*Who we are* as practitioners is always relevant, and nowhere more so than when we are engaged with, and affected by, issues that belong to all of us, and are likely to elicit strong feeling responses from us.

We will explore this systemic perspective further, together with some of the ways it can inform our therapeutic practice in addressing issues of climate crisis and justice. We'll share a brief description and examples of the contribution that Processwork theory and practice makes to awareness in this field and will offer examples of therapeutic interventions.

## Processwork[1]

Processwork emerged as a therapeutic modality in the 1980s, through the work of Dr Arnold Mindell, and colleagues, drawing on his practice as a Jungian analyst, as well as his previous study of quantum physics, and his deep curiosity about life processes. As a Jungian, Mindell recognised the capacity of our dream life to express the necessary parts of our wholeness which are less known and further from our identity. Processwork began to evolve further as a systemic awareness practice when Mindell recognised that this was true not only for night-time dreams, but also for our body

DOI: 10.4324/9781003436096-14

symptoms, relationship conflicts, accidents and synchronicities – in fact, to everything that is part of our experience but not our identity. 'Dreaming' is happening all the time. Hence what leads Processwork practice is the process itself: 'following nature'. It is inspired by, and has affinities with, theories and practices from many times and places, including Taoism, Gestalt, Aboriginal Dreamtime, Systems Theory, social justice movements and multi-dimensional understandings of the nature of reality.

Before exploring the dynamics of human interactions more fully, we acknowledge the context of the world of human and other-than-human beings. The following exercise is an invitation to locate ourselves within that wider context.

### *Arriving ...*

> *Imagine that you have just arrived to participate in a residential group with people you don't yet know. Before you sit in circle together to get to know each other, you are invited to go outside, find a bush, a rock or some aspect of the place that you feel drawn to, and introduce yourself there. To speak as fully as you might to a trusted friend, sharing how you are feeling, maybe what's on your mind for the group event, what you have just come from in the rest of your life, and so on. And when you have said all you want to say, the invitation is to stay and listen – with all of your senses, not only your ears – to whatever response you experience. And when that is done, to allow yourself to imagine becoming that 'other' that you've been conversing with and then noticing what is new or different about experiencing the world from here, and any insights or useful perceptions for you and your life – this moment, your path, your work, your relationships – that this brings. Finally, to return to your more familiar experience of self, take a slow moment to thank the place, and return to the group of humans.*

You might want to go outside and try this now, taking a few moments before you continue with the rest of this chapter. It is offered as an encouragement to remember and reconnect, using all of your senses, with your relationships with the other-than-human world. Thich Nhat Hanh (2017) put it beautifully:

> To 'be' is always to 'inter-be'. Our body is a community, and the trillions of non-human cells in our body are even more numerous than the human cells. There are no solitary beings. The whole planet is one giant, living, breathing cell, with all its working parts linked in symbiosis. We do not exist independently. We inter-are. Everything relies on everything else in the cosmos in order to manifest – whether a star, a cloud, a flower, a tree, or you and me.

Increasingly, voices in the global north are aligning with cultures that have never lost their understanding of our kinship within the circle of life. Mindell identifies nature and our world as *systems with awareness* (Mindell, 2010).

How can we, as therapists and facilitators, practise this systemic awareness that is a lived expression of ecology, to inter-be – in our work with clients, and in our own lives and relationships within the world? And how does that resource us in times of ecological and climate breakdown? This chapter explores some of these deep questions. On the level of human interaction, we examine some collective dynamics of power in both inner and outer worlds, and how these are relevant to our systemic awareness.

## Who We Are as Writers

Our own positionality here as authors is relevant because the lens we use to examine these issues will inevitably impact our perspectives, however much awareness we aim to bring. Noticing and being accountable for areas of unconsciousness around privilege, for instance, is an ongoing practice, not a one-off exercise.

We are two white cis-gendered women of diverse sexuality, middle class, with all the privileges this brings us living within the UK. As UK citizens, we have also been raised within its unprocessed colonial history and continue to benefit from that and are accountable in relation to it. We recognise the courage, huge work and dedication of many people within the UK and around the world to raise awareness about these structural dynamics, to bring them into environmental awareness and campaigning work every single day, as their and our lives depend on it.

Therapists from different modalities may feel encouraged to reflect and discuss with colleagues how to embed this awareness within their client practice. It is not a question of fostering guilt, but of bringing compassion to ourselves and others and using any privileges for the well-being of all. That is, 'all' on a human scale *and* an other-than-human scale.

## Collective Dynamics of Power

Climate and ecological breakdown are structured by dynamics of power and discrimination. These dynamics operate on every scale, from the global and collective, to the interpersonal and relational, and are internalised intrapsychically. Penniman (2023) observes: 'It stands to reason that any hope of solving the environmental crisis will require an examination and uprooting of the white supremacist ideologies that underpin the crisis.'

The issues that face our world, our clients, our beloveds and ourselves are challenging us to become more aware of the history of colonialism, extraction, systemic oppression and global injustice. In particular, in relation to the environment, we are challenged to become more aware of the devastating impact brought by the centralisation of power by capitalist culture – a culture which assumes dominance and entitlement to extract and take from the earth, from the peoples of the earth and from the spiritual cultures that have sustained them. The alienation this brings is beginning to be addressed more readily within dominant cultures around the world, reintroducing us to experiences and values of reciprocity, as Kimmerer (2013) has explored so beautifully.

Where this has not yet taken root, we are often complicit in patterns of supremacy and discrimination, without awareness or accountability. 'We' refers here especially, but not only, to those of us with privilege in the dominant culture. Those who are targets of discrimination also often suffer from internalised patterns of oppression. Patterns and behaviours of discrimination range from subtle micro-signals to overt and legally sanctioned aggressions and threats to life.

Here are some key things to notice in identifying structural discrimination:

*Structural discrimination* (SD) is embedded throughout society and its social institutions, and normalised in our attitudes, policies and practices. A key aspect of addressing SD is noticing how our signals, biases, actions and beliefs are conditioned and influenced by historical aspects of oppression. Simply put, SD is imbued and mirrored in our long-term behavioural patterns, attitudes and momentary signals.

*Internalised Oppression* is when we unwittingly internalise our external oppressor/oppressive system and its behaviour towards ourselves, adopt their style and belief systems and accept their criticism of ourselves. It's important to frame it clearly as a *symptom* of SD, rather than as an individual psychological problem. Seeing it as a personal psychological problem *only* is part of what perpetuates SD and exacerbates shame. Yet the way SD impacts us so painfully and so deeply, requires us to still work on our own experience of internalised oppression, which can be liberating and empowering.

*Internalised Supremacy* is also symptomatic of SD. Members of the dominant group are conditioned to see their positions as natural and earned, and to internalise a sense of being more deserving of the resources of society and the advantages associated with those resources. Internalised supremacy also marginalises parts of oneself (Karia, Amerasekera and Audergon, March 2022; unpublished).

## Rank and Privilege

Structural discrimination and privilege lead to dynamics of conflict that are not *symmetrical*: power is not equal on both sides. Processwork uses the term 'rank' to help identify signals of privilege in interactions. *Social rank* and privilege include our positionality according to racialised identity, ethnicity, religion, health, age, gender identity, sexual orientation, financial resources, class, education and more.

*Contextual rank* depends upon the environment we are in. We may have *psychological rank* from having internalised patterns of loving support, and/ or from having overcome and gained strength through difficult and traumatic life experiences. *Spiritual rank* may come from experiencing a deep connection with, and validation from, a greater and transcendent power.

Rank is related to our personal power, and usually the more we have, the less we tend to notice that we have it. It can be hard to be accountable for our high rank. Unconscious high rank not only affects interactions with others but has a profound impact on the mental and physical health and self-value of others and is a vehicle for internalised shame and oppression (Morin, 2014; Diamond, 2016).

## Trauma

In writing about structural discrimination, rank dynamics and trauma, we feel it's important to go slowly, and to stay connected, as writers and readers, with our embodied experience in the moment, as we bring our awareness to this field, recognising the enormity of pain and suffering there is from discrimination and trauma in this world. We need to recognise that we each come to the topic with a range of feelings including grief, anger, guilt, rage and fear, hope and hopelessness, a viscerally held history and current experience of discrimination – and sometimes the respite and sanctuary of dissociation and disconnection. We may also be grappling with questions about what we each might be able to contribute to the world.

Much has been researched, developed and written in recent years to enable us to become more skilled and aware in considering trauma, particularly as practitioners. (e.g., Menakem, 2021; Ogden, 2006; Fisher, 2021; Mate, 2022). We know more about the neurobiology of dissociation and splitting, the trauma responses of fight, flight, freeze and appease, how to work with flashbacks, and the range of ways in which we process acute and chronic traumatic experiences in our lives and the lives of our clients, including intergenerational impacts.

When we examine the interweaving of personal and collective aspects of ecological crisis and structural discrimination, and the way the world 'out there' and the world 'in here' are entirely interconnected, we can identify patterns of traumatic experience that have a similar impact at an individual and collective level. Arlene Audergon, writing in *The War Hotel* (2005), draws on personal and professional experiences across different parts of the world, including several years of working with traumatised and fractured communities in post-war Balkan states. She compares the symptoms of individual post-traumatic stress disorder as listed by the DSM-IV with signs of communal and collective trauma, and finds fundamental equivalence in a number of areas, including the persistent replaying of past traumatic events in the present moment; hypervigilance and arousal; reactivity which can be vulnerable to manipulation; detachment and disinterest; hopelessness and an inability to function.

Even if our personal lives seem to have little direct daily impact from the global consequences of climate and ecological crisis, we live within an atmosphere that is patterned by an increasing volume of distress. When this is intangible and diffuse, it is no less potent. On the contrary: we are 'in the soup' of it with less opportunity for awareness or agency.

## Processwork in Practice

The foundations of Processwork are deeply ecological. It is based on the understanding, often referred to as Deep Democracy (Mindell, 2017), that every part of a process has its place in the whole, including those parts we welcome and those we disavow. Whatever is disturbing us already contains the seeds of its own creative solution or evolution, and contains a contribution to the whole that is, at that moment, over the edge of our identity or awareness.

Following nature encourages us to follow a process as it emerges *in its own terms.* This is an embodied practice, using *sensory-grounded information* that we perceive and experience in differentiated channels: movement, sound and speech, visual and felt experiences. We include two more complex channels of awareness with their own characteristics and dynamics: relationships and the world channel. A signal that catches our attention in one or more of these channels indicates a moment in a dynamic process that is implicit but not yet unfolded – like a snapshot of a movie. In particular we explore incongruent signals which contrast with conscious intention and identity. As a process emerges, something that begins as a disturbing body symptom may bring awareness to a world issue, or a visual experience of nature can bring momentary meaning to a stuck relationship issue.

## An Example of a Process Unfolding Through Channels of Experience

The following example[2] demonstrates how the facilitator invites the client to follow the exact signal in the channel in which it first appears, and then as it evolves. The facilitator, at a certain point, embodies the more known experience, leaving the client free to experience more fully the part that is less familiar.

The context is an experiential seminar exploring body symptoms, discovering the ways in which world issues are entangled in our most personal experiences.

> *Paul, an environmental activist in his forties, describes a pain in his shoulders. This may be understood as caused by the stress of the last few years of his life of environmental campaigning. But the facilitator is interested in what might be emergent in the process, so asks Paul to describe more how he experiences the pain.*
>
> *He says the pain is a tightness, a sense of contraction in the muscles. He leans forward slightly in his body as he says this, and his head starts to lean*

*down. Noticing the movement, the facilitator asks him to amplify the tendency, and he responds quickly and easily to this suggestion, leaning forward more in his shoulders and head. He sighs and says it feels like a great weight is on his shoulders. This weight now becomes the lesser-known part of the process or system, so the facilitator offers to be the one leaning forward so Paul can embody the heavy weight.*

*He jumps up relieved and starts pushing down, becoming quite excited as he does so. He starts to become bigger in his posture, filling the space and enjoying it more. Suddenly an image comes to him and takes him by surprise: 'Oh wow!' he realises, 'I am the world!' He has an image of Atlas – the Greek mythological figure who carries the world on his shoulders. Suddenly he feels seen for his role in life and feels empowered by it. Becoming Atlas, and the world itself, he feels both parts together: the embodied stance of carrying the world and experiencing the strength and vitality of the world itself, through his body. Now he is no longer two parts, one oppressed by the other, but a unified experience of strength. From here, he has a pattern for using this energy in his activist work.*

On a systemic level the work shows how Paul had internalised an oppressive weight, an embodied experience that reflects collective marginalisation of the activist role. We could also say he experienced a low rank as an environmental activist. But through unfolding the process in his shoulders he came to experience a mythic dimension of the role that gave it global meaning, and enabled him to experience great strength, joy and a sense of his deep spiritual connection and rank. His identification shifted, and his experience of the way forward became clearer.

## Different Kinds of Awareness

We have different kinds of awareness experiences, and Processwork differentiates three in particular, with specific characteristics and inviting different facilitation.

*Consensus reality (CR)*: this refers to ordinary reality as we understand it in mainstream materialist culture. We more or less consent or agree on what is real, which can be measured and repeated: for example, the daily numbers of biodiversity and species loss, or requests for access to mental health services.

*Dreamland or dreaming*: refers to dreams, and dreamlike experiences; feelings and experiences that we can talk about but not verify; myths, polarisations and visions. In this dimension, our identifications are less fixed, more fluid; we can embody different parts of a process beyond our CR identity.

*The Essence level*: this is named in many ways in different cultures. For some it is God or Spirit or the multiverse; elsewhere it refers to the unified field, beyond the polarities of the other levels. In recent years Mindell has referred to the Process Mind (2010), indicating the unified awareness and purposefulness of the all-inclusive field of life itself.

## Inner Work

The following example[3] demonstrates several of the elements we have been describing throughout this chapter. This piece of work was unfolded as an 'inner work' where someone facilitates their own process. It shows how the process is unfolded through noticing signals in channels and simultaneously being aware of the process as it moves between different levels of awareness.

This same practice is applied to working with processes at all scales: our own inner work, an individual client, a relationship, group or organisation.

*I'd woken and read that there was more flooding in my area due to heavy rains and that an oil company boss had been announced as the head of the COP28. I noticed my deep sense of hopelessness, like a shadow. Not quite sure how much attention to pay it, and not always sure what to do with it, I pulled myself together and headed off to a group supervision day.*

*The group decided to work together on the topic of climate collapse, however I was still left afterwards with this feeling of hopelessness, which was fast becoming overwhelming.*

*Later, I was meant to be facilitating a client in the group, and yet I was still totally overwhelmed. My supervisor suggested that instead of 'getting on and pulling myself together' that I take seriously what was happening and unfold the experience through an inner work.*

*The invitation to take my overwhelm seriously already changes something in the field and in me – this signals to me that my overwhelm and the associated feelings are important. It helps me as I begin facilitating myself in the experience.*

*Taking care of this feeling of overwhelm allows me to slow down. I bring my hands to my heart. I am with my body sensations and alongside that part of me that is overwhelmed. Two parts: a part that is overwhelmed and a part that can take care.*

*As I do this, my breathing begins to slow down, a still spaciousness opens up inside me, in my belly. I want to move back in my chair. I move back so far that I lose the quality. I come forward again and re-find this still spacious quality. Within the quality I notice a pulse inside my belly and bring more awareness to this by amplifying the movement in my hands and arms.*

*I have the impulse to stand – I know that this might help me experience the movement unfolding more fully. In doing so, I notice two parts again – something moving downwards through my legs, and something rising through my torso. It is this second part that surprises me most – I hadn't been expecting that. So, I pay attention to this experience, following the internal movement sensation of something rising inside me. An image appears of a waterfall falling upwards to the sky. From the earth to the sky.*

*I stay noticing this internal movement quality and then bring this quality into external movement. As I stay with this quality I notice its essence of light, effortless rising. The moment is full of joy.*

*Through unfolding my experience of overwhelm and hopelessness, I find this new emergent quality of effortless rising. From this less well-known quality I ask myself, where do I need this quality in my life, how could this quality help me in my engagement with climate collapse?*

*I notice that I don't have to try so hard to change things, that I can bring this quality into my facilitation and be alongside these issues with a deep connection to the essence dimension.*

There is a connection between the feeling of hopelessness and the collective context: it comes from a worldview that one should be 'up' or productive. That is a central characteristic of the wider systemic problem of climate change – the dominant intention of trying to be more productive whilst marginalising the body and nature. By unfolding it as a sensory experience, the wisdom of the body becomes available. This work shows us how the worldview that pushes the climate to collapse is mirrored in the person's internalised process. When we address this dynamic individually, this also has direct relevance to the wider collective. This work is vital when engaging in the complexity of environment and climate change.

## In Conclusion

We all need a community of colleagues and friends to support each other as we each find our own way to navigate the perils, distress, opportunities and the wonder of being practitioners and being alive in these times.

Whether we are aware of it or not, we are all in this living system of the world, with its hugeness, vastness and complexity. We are part of nature, we can only inter-be. Recognising this can evoke an increased impetus to take accountability for how we use our power. But also, to notice those subtle interdependencies within ourselves, between each other, and with everything outside of our awareness. This is the path of wholeness – like nature itself, this is the path that builds long-term sustainability and resilience.

## Notes

1 Links for more information on Processwork:
www.processworkuk.org
www.aamindell.net
Audergon, A. (2004) Collective Trauma: The Nightmare of History, in *Psychotherapy and Politics International*, 2(1). London: Whurr.

Audergon, A (2008) Daring to dream', in: B. Hart (Ed.) Peace-building in Traumatized Societies. Lanham, MD: University Press of America.
Audergon, A ( 2006) Hot Spots: Post-Conflict Trauma And Transformation, *Critical Half* 4(1). Women for Women International.

2 Permission was given for both examples of client work given in this chapter. The written records have been seen and agreed.

3 Workshop participant material anonymised to preserve anonymity.

## References


Audergon, A. (2005) *The War Hotel: Psychological Dynamics in Violent Conflict*. London: Whurr Publishers Ltd.

Audergon, A. Amerasekera, E. and Karia, A. (2022) Metaskills and Some Key Concepts. *Unpublished Handout* from Processwork UK training seminar: Collective Trauma: The Facilitator's Path, Working with Dynamics of Trauma and Structural Discrimination.

Diamond, J. (2016) *Power – A Users Guide*. Santa Fe: Belly Song Press.

Fisher, J. (2021) *Transforming the Living Legacy of Trauma*. Eau Claire, WI: PESI Publishing & Media.

Kimmerer, R. W. (2013) *Braiding Sweetgrass: Indigenous Wisdom, Scientific Knowledge and the Teachings of Plants*. Minneapolis: Milkweed Editions Penguin.

Mate, G. (2022) *The Myth of Normal*. Weymouth: Vermilion Ltd.

Menakem, R. (2021) *My Grandmother's Hands*. UK: Penguin Books Ltd.

Mindell, A. (2010) *Processmind: A User's Guide to Connecting with the Mind of God*. Wheaton, IL: Quest Books.

Mindell, A. (2017) *Conflict: Phases, Forums, and Solutions*. CreateSpace Independent Publishing Platform.

Morin, P. (2014) *Health in Sickness – Sickness in Health: Towards a New Process Oriented Medicine*. Portland: Deep Democracy Exchange.

Nhat Hanh, T. (2017) *The Art of Living: Mindful Techniques for Peaceful Living from One of the World's Most Revered Spiritual Leaders*. London: Rider Publications.

Ogden, P. (2006) *Trauma and the Body: A Sensorimotor Approach to Psychotherapy*. London: W. W. Norton & Company.

Penniman, L. (2023) *Black Earth Wisdom: Soulful Conversations with Black Environmentalists*. Amistad: Harper Collins Ltd.

Chapter 7

# Rehearsing Radical Care: Motherhood in a Climate Crisis

*Celia Turley and Jo McAndrews*

## A Note from the Authors

The chapter that follows outlines the growing public attention to reproductive decision making in the face of the climate crisis and describes some of the values, approaches and learning which shaped a creative response. Whilst this chapter is authored by Celia Turley and Jo McAndrews, all the women who indelibly shaped the project are housed within our words. Colleagues Sophia Cheng, Liz Mytton and Maria Fannin are essential collaborators. The stories of the women[1] who courageously stepped into the creative process with us: Anna, Daniela, Jodi, Rosanna and Ruby, are the lifeblood of this work. Limited space means their creative monologues are not included. This chapter must therefore be understood as an accompaniment to their stories. Videos of their performances can be viewed on the Motherhood in A Climate Crisis webpage, and we urge you to watch them for an insight into the depth and range of feeling this topic reveals.

## The Context of Motherhood in a Climate Crisis

Motherhood in a Climate Crisis is a care-filled, collaborative theatre-making process that gives voice to frank, personal stories on how the climate crisis shapes if, and how, we mother. The project aims to create a consensual, non-judgemental space for women with diverse experiences to clarify and creatively communicate their thoughts, feelings and choices on parenting in an era of ecological uncertainty.

In 2022, the project's first iteration, we devised a six-week online and in-person workshop process with a group of women, which led to an intimate public sharing of a co-created performance and a series of filmed theatre monologues, available online. This public culmination was designed to express our stories to others, and confidently own them for ourselves. We hoped this might be a starting point for people to feel and think anew about the pervasive and unexpected impacts of climate change. We wanted to

DOI: 10.4324/9781003436096-15

stimulate courageous conversations amongst audiences about how their own experiences might be reflected – or not – in the stories they observed. We aimed to kickstart a public recognition of the decisions that arise as we hold the realities of climate change in our bodies, heads and hearts. We wanted to communicate the truth that climate change is not something which happens 'over there', in geographically and temporally distant realities. We are living with climate change now. It is impacting our emotional worlds, and for some, our most intimate bodily choices.

For many women, the existential fear of a future ravaged by climate breakdown adds uncertainty and stress to the decision of whether to have a child, compounding anxiety around an already heavily loaded question (Kamalamani, 2016). Western society's gendered expectations run deep; starting a family is still presented as the pinnacle of joy, fulfilment and meaning for women. Ambivalence about motherhood remains a powerful taboo. All genders may face questions about parenting, yet dominant heteronormative gender standards place the burden of responsible reproductive choice at women's feet (Agarwal, 2021). The pressure of the body-clock ticking is 'a societal and cultural trope' (Agarwal, 2021: 7) impacting reproductive choices. Compound the uncertainty of a fertility timeline with the countdown to irreversible ecological breakdown and there is powerful potential for overwhelm, inhibiting open communication.

There have been moments where this largely unvocalised disquiet has emerged in public conversation, such as Alexandria Ocasio-Cortez' questioning the 'legitimacy' of having a child in a potentially climate-ravaged future (Ocasio Cortez [@ocasio2018], 2019). Academic research has begun to respond to this nascent question. A pioneering 2020 survey of 607 US-Americans who were factoring climate change into their reproductive choices found 96.5% of respondents were 'extremely' or 'very concerned' with the climate impacts that their existing, expected or hypothetical children might experience (Schneider-Mayerson and Ling, 2020). A study which polled more than 10,000 young people across 10 countries found that four in ten were hesitant to have children as a result of the climate crisis (Hickman et al., 2021).

The project came to life at a time when the scale and urgency of the climate crisis had reached new depths of awareness within the UK, and climate activism was burgeoning. An important precursor to the project was Birthstrike, the 2019 upsurge of women and men who publicly declared their refusal to have children in response to the coming 'climate breakdown and civilisation collapse' (Pepino, quoted in Hunt, 2019). Movement activist Blythe Pepino described Birthstrike as a 'radical acknowledgment' of how the looming existential threat is 'altering the way we imagine our future' (ibid.). Mobilising the language of the political strike, the campaign highlighted the fallacy of dichotomy between our public, civic lives, and our personal reproductive decision making. Government failure to mitigate the climate crisis, Birthstrike argued, restricts bodily autonomy now and denies safety to potential future children. Yet within the media there was a recurrent framing of the campaign as one about the carbon impact of children, rather than systemic injustice and collective responsibility for the future of all life. This rested on assumptions about the child as consumer, obscuring the vast discrepancies in consumption levels that depend on birth location, culture and levels of privilege and wealth (Gaitán Johannesson, 2022).

These responses echoed a deeply problematic populationist rhetoric, directly connected to complex, brutal histories of racialised harm and creeping eugenics in contemporary climate conversation (Tilley and Ajl, 2022). The consistent eschewing of these complexities led to Birthstrike ending its public campaign, reshaping into a solidarity and support group, Grieving Parenthood in the Climate Crisis: Channelling Loss into Climate Justice.

This chronology illustrates the sensitivities to this topic and explains why, even with the growth of this conversation in the public realm, project founder Sophia Cheng identified the need for the Motherhood in A Climate Crisis project as being rooted in addressing a 'social silence': a sense of taboo, rather than total void of conversation. Later, we identified this as shame.

## The Birth of the Project

The genesis of the project was an initial one-off discussion hosted by Sophia Cheng within an online community of women. Sophia shared her then-ambivalence towards motherhood. A co-facilitator shared their story and the story of a mother's reckoning with climate change was read aloud. More women than usual turned up and a 75-minute session ran to three hours as women increasingly wanted to share their stories. Sophia realised she had tapped into something powerful; event participants felt it too.

> *It made me realise how lonely I felt about the topic before and that I am not alone with all those mixed feelings, hearing about your journeys. That was certainly an afternoon that I will draw inspiration from for a very long time.*

> *I just want to say thank you! What a group of brave and brilliant women you are! I felt connected and understood and that's something I will cherish forever.*

Sophia wrote a concept note immediately after the event and with Creative Producer Celia Turley developed a proposal. Could the polarising logic framing the issue be shifted if we asked women to join a non-judgemental exchange about feelings, not opinions, focusing on courageous conversations? This required participants to be in a brave space (Arao and Clemens, 2013), a group process that was compassionate and carefully tended, with recognition that the topic might move into painful territory. We understood that human stories are more powerful than data at communicating climate change. Stories can touch us profoundly: 'emotions are the gas you can't smell or see, yet it fuels public engagement' (Wray 2022). Could such a project not only help affirm feelings but also create new solidarities, allowing for a different kind of action on climate change?

## Creativity and Care

Creative storytelling projects and the use of narrative in social justice organising run the risk of reproducing the extractive logics of capitalism. We were driven by wanting to create a counter-model to projects we'd witnessed that took stories from participants for public consumption without attention to mutual support and time for creative exchange. As two not-yet-mothers in their mid-30s, Celia and Sophia were importantly being led not just by their professional expertise but their deep curiosity about other women's experiences and an intuitive sense of what might be needed. They recognised the tenderness of these topics and wanted to centre care in the organising practices.

They discussed their ideas with the Climate Psychology Alliance and invited Jo McAndrews into the project as Therapeutic Facilitator. Whilst Celia and Sophia knew the importance of this role, they recognised deep skill was required in meeting murky, unspoken aspects of human experience, and wanted guidance. This came from Jo's wisdom in nurturing resilience in parents and families to face the realities of climate change. Liz Mytton joined as Creative Facilitator; her rich and varied experience working on collaborative theatre projects was vital for leading the artistic direction of the work. Liz identified that she and Jo would need supervision to work through some of their own responses to the project. This led to one way we tried to embed a culture of care into the project, by setting a budget for therapeutic supervision. We were also accompanied in the project by Professor Maria Fannin, specialising in feminist perspectives on Reproductive Health.

## Setbacks Around Missing Voices

Our intention was to include a wide range of voices and experiences. Many conversations about the climate crisis and on motherhood centre white experience and exclude the voices of people of colour. This leads to a skewed narrative that reinforces dominant racist power structures. The team was predominantly white, with Liz the only black member. We did our best to be conscious together of how racial differences play out in our culture, but we did not always get it right. In the invitational publicity we tried to reach a racially diverse audience, but attracted a group of all white women, albeit with other diverse identities. Reflections on how the topic might be experienced across race and ethnicity were therefore missed.

We had failed to take time to build the trusting, organising coalitions essential to ensure that outside of the organising team, women of colour were actively involved. The workshops were intense and time-heavy, without payment, which may have excluded those without access to financial and family support, though childcare and transport costs were provided. We considered delaying the whole project, not wanting to present stories of only white perspectives. However, those in the group had already committed a lot of time and energy to the project and we wanted to honour that. We decided to make it clear the first performance was an introductory sharing and not a definitive exploration. It would be an opportunity for us to learn and build the next phase of the project.

There were very different experiences in the group, including women who were birth mothers, adoptive mothers, ambivalent about motherhood, feeling unable to justify becoming a mother, making the decision not to have a second child, grieving having had children before becoming aware of the climate crisis, women who were heterosexual and queer, single and in partnerships.

## Telling Our Stories

The project took the form of a series of full weekend workshops in Bristol and evening sessions on Zoom. Liz designed a programme of creative, embodied, courageous story and theatre practices. We played games, made theatrical tableaux, told stories, discussed climate science, wrote radical new words to old songs, composed poetry and blessings, moved as a chorus to express emotions and togetherness. There was laughter, rage, tears and group support. The creative process allowed depth of connection and expression. It gave us ways to speak the unspeakable, break taboos, and to form a deep, shared experience of courage and care.

> *I think this was the only true safe space I have ever been in.*
>
> (Ruby, project participant)

Liz helped the group write a script of the story they wanted to tell publicly, arising from the conversations and creative experiences of the workshops. She facilitated a collaborative process to bring it to life in simple physical, visual and musical ways to support each person to tell their story in an informal 'open rehearsal' for friends and family members. Each story was then recorded by a professional filmmaking crew, along with some interviews with project members, so that we now have a series of stories to share and build on as the project develops to include more voices.

> *I'm proud of how as a group of women we thrashed out some big topics and supported each other throughout. The richness of the project means that I am not entirely the same person I was when I joined.*
>
> (Anna, project participant)

The combination of creative collaborative exploration, deep care and climate justice activism was staggeringly powerful.

> *If my only contribution to this project is to help expand what being a mother means, then I will be eternally happy. Being a mother is not an option between giving birth, and not giving birth. It's a call for courage and fierce love in a world where conditions are unfavourable. You can mother your mother. You can mother a sibling. You can mother a child that did not grow in your belly. You can mother the earth. For me, the question isn't to mother or not, but how to mother. It's also important to me that I allow myself the freedom to change my mind, to tear up the transcript and start again. We make every decision we have with the best of intentions, armed with the knowledge at our disposal and the feelings in our belly. When I go from a twenty-something to thirty-something my feelings towards motherhood may change, but my desire to protect this Earth and ecosystem never will.*
>
> (Jodi, project participant)

## A Culture of Care

The team recognised that it was vital to hold with deep care the women who joined us in sharing their stories, longings, grief, regrets, pain and love. The most basic expectation was that the therapeutic practitioner would be there as a safety net, someone to go to for anyone who felt distress and in need of support. Knowing that a therapist was on the team helped everyone feel safer to go deep with their experiences. We could all feel confident that if anyone fell apart there would be someone to catch them.

However, we had a bigger goal than that: to provide such strong care in every part of the project, that being involved in it, including as audience, would be a nourishing, inspiring experience. In practice it is very rare for any project to be this explicit about the need for care throughout. We were more familiar with the idea of having someone therapeutically trained to pick up the pieces when people became distressed. Yet we wanted to learn how to work in such a relational, inclusive way that distress was minimised. And more, we hoped a care-centred process could offer us an experience of connection so affirming that it would show us new ways of creating a culture that grows relational resilience in the face of this unfolding crisis. None of us had ever worked like this before and we were excited to see what we would learn. Below we distil some of the central tenets in this experimental practice.

## Key Elements of Our Care Practice

### *Relational Welcome and Warmth*

This is the understanding that a resonant relationship is at the heart of wellbeing and resilience (Peyton, 2017). We understand that humans do not function as individuals; everything we feel and do is in the context of our relationships and how much we feel we belong. Our nervous system states are infectious; if there is distress or anxiety in the room, we will feel it – likewise safety and warmth.

### *Active Welcoming of Difference*

We invited women to bring a range of experiences of motherhood, including the decision not to give birth. These are deeply emotional issues carrying the potential for judgement and feeling vulnerable and isolated in positions different from others in the group. We modelled and invited a complexity of experience.

### *Radical Listening*

This is an approach to listening in which the listener welcomes whatever the speaker wants to express in a non-judgemental way, without comment and with empathic, resonant reflection (Peyton, 2017). The aim is to provide an atmosphere of warm curiosity, and to meet the basic human need for being seen and heard, understood and welcomed. It is radical because it is different from dominant Western experience, where listening comes with opinion, attempts to solve, advise or facilitate the speaker into something more valuable to the listener. Working with embodied responses using theatre and movement practices enabled deeper understanding and experience.

### *Trauma Awareness*

Trauma is complex in its origins and is not just about overwhelming events that have happened to us. It is also collective and cultural. When we grow up in a family and culture that does not meet our needs for safety, our nervous systems become traumatised. This includes growing up with inequality and lack of inclusion, such as systemic racism, as well as gender, disability, neurodivergence, age, sexuality and other socially constructed discrimination (Menakem, 2017). A trauma-sensitive approach is mindful of everyone's need for autonomy, to regulate their need for safety in the way that makes sense to them. It means always offering choice and checking for consent at each step of the process.

### *Anti-racist, Feminist and Other Anti-discriminatory Practice*

This is recognising and being willing to question the models of white supremacist capitalist patriarchy in Western cultures (bell hooks, 2014). These systems create a culture of separation and domination that is anti-care. We recognise that we have been conditioned by these systems and will unconsciously perpetuate them unless we actively learn to do something different (Kashtan, 2014).

### *Shame Awareness*

Shame is always present in group processes and is destructive when unacknowledged. A shame state is hugely distressing to the person experiencing it and can cause them to withdraw or attack in an attempt to regain safety. Shame is dissipated in the presence of warm support and acknowledgement. In Western patriarchal cultures, shame is used to control children and is a common experience for women. In a project inviting women to share publicly experiences around motherhood, it was certain that shame would emerge. When we understand its dynamics, we can meet it with support and understanding.

### *Climate Psychology*

Engaging with the climate crisis in any way will invite feelings of despair, anger, fear, grief, numbness, betrayal. We invited women to engage with the hugely personal and emotional subject of motherhood in this context. It was important to offer a strong practice of care for the deep feelings evoked by this. Climate distress can be relieved by acknowledgement, support and collective organising. Women were drawn to the project as a form of action, to communicate the urgency and impact of climate change.

### *Welcoming Neurodivergence*

Neurodivergence is an influential factor in group processes. When it is not named or acknowledged, there is an assumption of neurotypical processing and behaviour, which can create distress and conflict. When neurodivergence is actively welcomed, we can allow for different needs and experiences, and welcome the whole person. Both the therapeutic facilitator and the creative facilitator are neurodivergent. Welcoming this in the group and being explicit about needs allowed for a radical and rare sense of safety and acceptance.

### *Playfulness*

Our creative process used games to encourage communication, reconnection with our imaginations and ability to play. It helped us see each other as creative, playful beings and quickly remove barriers such as perceived hierarchies, and normative structures. Games also serve a purpose as an environment for shaping and testing ideas. They have value in discharging stress, inviting laughter, facilitating moments of physical and emotional connection and bonding, pushing boundaries and experimenting.

## Challenges to a Culture of Care

There were significant obstacles to centring care. It was easier to ensure that we supported the women we worked with, the 'clients' in the project, than it was to bring care into the way we worked together as a team. We each had assumptions about what care means and have internalised the uncare of the cultures we live in. At first, we did not address those assumptions; we imagined that we had a shared understanding. Some of us feared that care meant having to endlessly talk about feelings or had difficult previous experiences of therapeutic work. Having a specific care practitioner at times led to outsourcing responsibility for care and imagining it inhibited efficiency.

The team was so used to living in a culture that devalues feelings, listening and the need to belong that it was sometimes hard to tend to relationships and inclusion. During the planning phase we unconsciously replicated the capitalist productivity culture of focusing on tasks and deadlines. There was much to organise and little time to explore what was really needed to build care throughout. This led to tensions later and members of the team feeling isolated and confused. Some issues echoed the themes of both motherhood and climate crisis: insufficient time, inability to do 'enough', neglecting needs; an overwhelming sense of urgency and helplessness.

## Themes and Experiences Emerging Through the Sharing of Stories

Our stories and conversations brought up many complex and interconnected themes. They include ambivalence about becoming a parent, feelings of confusion, regret, rage, grief and shame, a sense of taboo and being silenced, dismissal both from others and within oneself and a sense of inadequacy and complexity. There was also internal and external conflict, pain in difference, awareness of aspects of injustice and privilege, honouring our differences and developing compassion for ourselves and others and palpable relief in sharing the experience.

## Applying Our Learning

We acknowledge that many reading this chapter are not looking to organise a creative project. We share our learning as we believe it is relevant to everyone involved in the support of climate aware mental health, in whatever capacity. The urgency of the climate crisis requires us to question our frameworks and models of working, especially when they have been created within the dominant systems of power and separation. We would love to see more informal and community approaches to mental health and wellbeing so that support is not limited to those who are able to access professional help.

Participants told us that the courage and informality of the group allowed them to share things they have never spoken about before. One member had believed that she would not be able to speak of her fierce grief that having a baby felt so irresponsible that it was impossible for her to consider. Her strength of feeling made it very dangerous to share; she had felt isolated in conversations with other women.

> *That room was the first time I was able to talk to people who were already mothers about how I truly felt. The subject has always felt impossible to bring up, even with other childless women, because you never know who will announce their pregnancy next and who you are going to upset. It is, in my opinion, the most deeply personal decision you are ever going to make. This project has been the safest and most supportive space I could have hoped for, which made discussing these unbelievably personal issues possible and has changed my life forever.*
>
> (Ruby, project participant)

The complexity of feeling and experience around having children at this time of crisis is much more widespread than is discussed both in everyday culture and in professional mental health support. It is a huge and painful subject that can feel dangerous to voice in a culture already judgemental about motherhood and denying of climate change. This makes it important that practitioners find support for their own complex feelings and thoughts about the subject, in order to hold their clients' experiences.

There is confusion and conflation in the public narrative about the question of population size and a racialised narrative that blames population numbers rather than inequality of consumption. We recommend critically engaging with this issue to locate and identify your own thoughts.

We invite you to use the videos as a resource in training and individual work. The power of hearing others' personal stories can start conversations, acknowledge the complex layers of personal experience and help people realise they are not alone. We decided to start with inviting people who identify as women, but this is also an issue for those who identify as men, non-binary people, and can also be a strong tension in couples. Young people approaching or within childbearing age will likely be under internal and external pressure to think about whether they want children. Social and family expectations can intensify this pressure and deny support. We think it is important that we do not wait for this to emerge on its own. If we want to be clear that we welcome our clients' experience around parenting in a climate crisis we need to find ways to demonstrate that proactively, even bringing it up directly.

Time is a vital resource which is often unacknowledged. 'Moving at the speed of trust' (Maree Brown, 2017) often means moving slowly! Time is at work in a very fraught way within the climate crisis. There is an urgent need for action, and also a vast and incremental geological timescale at play, which can make things feel intangible and beyond human influence. Both can lead to overwhelm and will likely impact work on this topic, through enactment of dominant modes of 'productivity culture' but also subconsciously. We recommend being alert to this and how it plays out in your own climate response.

## The Privilege of Choice

Whilst we achieved a breadth of experience in relationship to motherhood within the project, there was an underlying reality of a degree of reproductive choice within the group. Sistersong, the women of colour collective who coined the term 'Reproductive Justice' in 1994, developed the definition as 'the human right to maintain personal bodily autonomy, have children, not have children, and parent the children we have in safe and sustainable communities' (Sistersong, 2023). Key to their formulation is the idea that 'There is no choice where there is no access' (ibid.).

During the first phase of the Motherhood in A Climate Crisis project, the US Supreme Court ruled to overturn *Roe vs Wade*, ending the constitutional right to abortion in the US. In the UK where there is still free access to reproductive and maternal health care, huge disparities in provision remain. A 2021 report published by MBRRACE-UK found Black women in the UK were four times more likely to die in childbirth than white women (Knight et al., 2021). As assertions about the value of 'life' are mobilised by the political right for regressive limits to the lives of many women, there is an onus on climate movements to highlight the connections between reproductive and climate justice and articulate a vision of sustaining earthly life that leaves no-one behind.

## Last Words

The Motherhood in A Climate Crisis project is an experiment in finding our voices and practising care as a portal to different kinds of conversation about how the climate crisis feels and is mediated through our bodies. Artistic co-creation, which recognises that what is learned, felt and shared in the process is as vital as any final product, was a powerful tool in helping us reach the layers of experience that exist below the surface.

> *Using theatre and using the writing has helped me to develop a story that I didn't know was fully there.*
>
> (Anna, project participant)

An ethos and practice of radical care is the supportive infrastructure that made this possible. It didn't mean that we avoided replicating systemic power dynamics and societal failings. But when we inevitably experienced challenges, we aimed to address them, operating with the possibility of repair and change. This offered something vital: the opportunity to practise ways of 'being together otherwise' (Van Heeswijk, 2021). As we allowed for creative spontaneity, tended to uncertainty, communed in our grief and developed an ethos of radical care, we also rehearsed the values and embodied the practices that we need to live with joy and justice in climate changed futures. We hope this provides models for ways of being that expand into all aspects of our lives as we resource ourselves in a changing climate and changing world.

## Note

1 All participants gave consent to audio-visual recording of their performances and to the project being described in articles available both in print and in e-versions. They have agreed to the quotes used within this text.

## References

Agarwal, P. (2021) *(M)otherhood: On the Choices of Being a Woman.* Edinburgh: Canongate.

Arao, B. and Clemens, K. (2013) *From Safe Spaces to Brave Spaces: A New Way to Frame Dialogue Around Diversity and Social Justice in The Art of Effective Facilitation.* Stylus Publishing, LLC.

Gaitán Johannesson, J. (2022) *The Nerves and Their Endings Essays on Crisis and Response.* London: Scribe Publishing.

Hickman, C., Marks, E., Pihkala, P. et al. (2021) Climate anxiety in children and young people and their beliefs about government responses to climate change: A global survey. *Lancet Planetary Health*, 5(12).

hooks, b. (2014) *Teaching to Transgress: Education as the Practice of Freedom.* London: Routledge.

Hunt, E (2019) *BirthStrikers: Meet the Women Who Refuse to Have Children Until Climate Change Ends.* The Guardian. 12/03/2019. Available at: www.theguardian.com/lifeandstyle/2019/mar/12/birthstrikers-meet-the-women-who-refuse-to-have-children-until-climate-change-ends [Accessed 27 June 2023].

Kamalamani (2016) *Other Than Mother – Choosing Childlessness with Life in Mind: A Private Decision with Global Consequences.* Hants, UK: Earth Books.

Kashtan, M. (2014) *The Fearless Heart: Inspiration and Tools for Creating the Future We Want. Courage to Live it Now.* Available at: https://thefearlessheart.org [Accessed 4 July 2023].

Knight, M., Bunch, K., Tuffnell, D. et al. (eds) on behalf of MBRRACE-UK (2021) *Saving Lives, Improving Mothers' Care – Lessons learned to inform maternity care from the UK and Ireland Confidential Enquiries into Maternal Deaths and Morbidity 2017–19.* Oxford: National Perinatal Epidemiology Unit, University of Oxford.

Maree Brown, A. (2017) *Emergent Strategy: Shaping Change, Changing Worlds.* Edinburgh: AK Press.

Menakem, R. (2017) *My Grandmother's Hands: Racialized Trauma and the Pathway to Mending our Hearts and Bodies.* Las Vegas: Central Recovery Press.

Ocasio Cortez, A. [@ocasio2018] (2019) Alexandria Ocasio-Cortez says it is 'legitimate' for people to not want children because of climate change Instagram 26/02/2019 Available at URL: https://cdn.jwplayer.com/previews/4l2nM2AT [Accessed 1 December 2021].
Peyton, S. (2017) *Your Resonant Self: Guided Meditations and Exercises to Engage Your Brain's Capacity for Healing*. New York: W. W. Norton & Company.
Schneider-Mayerson, M. and Leong, K. L. (2020) Eco-reproductive Concerns in the Age of Climate Change. *Climatic Change*, 163: 1007–1023. 10.1007/s10584-020-02923-y.
Sistersong: Women of Colour Reproductive Justice Collective (2023) *About Us*. Available at: www.sistersong.net/about-x2 [Accessed 30 May 2023].
Tilley, L. and Ajl, M. (2022) Eco-socialism Will Be Anti-Eugenic or It Will Be Nothing: Towards Equal Exchange and the End of Population. *Politics* 43: 2.
Van Heeswijk, J., Hlavajova, M. and Rakes, R. (eds) (2021) *Toward the Not-Yet: Art as Public Practice*. Utrecht: BAK, basis voor actuele kunst.
Wray, B. (2022). *Generation Dread: Finding Purpose in an Age of Climate Crisis*. New York: Knopf Canada.

Voice 6

# Becoming an activist parent

*Chloe Naldrett*

On a bright sunny morning in February 2019, it hit me. Out of nowhere. All of a sudden, as I stood in the shower, I was staring into a terrifying and all-too-possible future. One where I was unable to feed my children, unable to keep them safe.

I have two boys, aged 13 and 10 as I write this. I brought them consciously and lovingly into the world, both deeply wanted and cherished from the moment of their conception, and suddenly I was imagining whether I might ever have to consciously author their exit from it.

I don't write this lightly.

The only comfort in the face of this horror was, and is, connection. Shortly after this I made a resolution to step into the environmental movement. I started listening and learning. I took my boys to the October 2019 Rebellion in London for the day: they held their own signs at the foot of Nelson's Column in the rain. We spoke to people, who were kind and friendly and grateful to us for being there. In September 2020 I sat, naked to the waist, outside Parliament, baring the breasts with which I had fed my sons as part of an action highlighting the vulnerability of women and the media's censorship of the climate crisis. I was arrested for the first time.

I've subsequently joined actions which have resulted in arrest four further times, including spending a week on remand in Foston Hall after breaking an injunction at an oil terminal in 2022. I've written and given talks to my family, friends, neighbours, and colleagues, and in schools and churches. I've become a spokesperson for Just Stop Oil, speaking to journalists on the radio, on zoom and in the studio. I never imagined myself capable of these things, but fear is a powerful motivator. Every one of these actions is an attempt to outrun the fear seeded in me by that awful vision.

These actions have cost me friends, and at times threatened my relationships with my family. They have cost me a certain amount of my privacy. The grief of losing everything that we love, and our hope for the future is crippling at times. It adds a further dimension to the challenges of parenting children, as my children's wants and expectations clash with my values and lifestyle in ways that they find restrictive.

DOI: 10.4324/9781003436096-16

But in 2023, with the Climate Crisis unfolding in plain sight all around us, it is not enough to say that we love our children. It is our duty as parents to fight for their future. Anything that I can do which pushes them further away from the fire, anything I can do which lessens the burden of responsibility which will fall to them, I will do. This is what I mean when I say I will do anything for my kids.

Tuning into the emotional and physical reality of the Climate Crisis has provoked a permanent shift in mindset which has irreversibly changed every aspect of my life. I can't imagine talking to a therapist who did not understand what it means to cross that threshold, or who did not know how to hold the range of emotions which have been opened in me. To have these emotions recognised and validated, not minimised, or soothed away, is the essential work required by activists.

Chloe Naldrett, UK

Section Three

# Becoming a Climate Aware Therapist

Chapter 8

# Climate Aware Therapy with Children and Young People to Navigate the Climate and Ecological Crisis[1]

*Caroline Hickman*

> *Tell me when it will stop hurting? Tell me when I won't be scared all the time. But you can't, can you.*
>
> (Locci, aged 12, project participant)

This is the problem we have when thinking about therapy in relation to the climate and biodiversity crisis, we know that the hurt and fear won't stop, it cannot stop, because the crises themselves will not stop, not now. We are not at the start of the story about these crises, we are halfway through the story, and still so often continuing to act as if the crises were not that bad, that there is plenty of time to resolve things, that there can be a happy ending to this story, as in all the fairy stories and films that we prefer children to watch in the Western world to give them the message that it will all end happily. So, we must ask the question, how do we talk with children about this without lying to them, minimising things, greenwashing, pretending, denying, yet without traumatising and terrifying them. These are important dilemmas.

If we accept the argument that anxiety and distress are mentally healthy and understandable responses to climate change and ecological emergencies then we can validate children's distress, tell them that they make sense, that we understand that they are sad and that they are not alone with their anxieties – all good relational responses (McAndrews, 2023). We can acknowledge that it is emotionally and cognitively distressing and hurts (Marks and Hickman, 2023). But then we are left with other questions; should we be supporting children with individual therapy when it is a relational systemic problem and moral injury (Weintrobe, 2021), does this risk redirecting the 'problem' back into the individual, and how do we offer support without falling into our need to 'ease' or 'remove' or 'reduce' children's distress or even to reduce our own guilt, grief and shame? How can we pay attention to individual distress within an ecological frame, hold the tension they create and not try to make these dilemmas go away, but sit with them and stay aware of these tensions?

DOI: 10.4324/9781003436096-18

> *You need to tell us the truth, because if you don't tell us the truth you're lying to us, and if you lie to us, we can't trust you, and if we can't trust you, we can't tell you how we feel, and then we're alone. But don't tell us the bad news all at once, tell us some bad news then some good news, then some bad news and then some good news. And anyway … I'm not a baby.*
>
> (Sophia, aged 8, project participant)

Research tells us about the impact of the climate and biodiversity crises on children and young people worldwide, examining how it affects children emotionally and cognitively, in their daily lives, and their trust and faith in government action (Hickman and Marks, 2021): 67% agreed that 'climate change makes me feel sad and afraid', and 62% felt anxious. We might reassure ourselves that children in Europe are currently somewhat protected from the impact on daily living, with only 26% in the UK, 31% in Finland and 35% in France identifying that it impacts going to school, spending time with friends or in nature and studying, compared to those in the Philippines and India (74%), not that we are reassured by this global injustice. But when we look at the figures showing cognitive impact, we can see how children globally may have more in common with each other than difference. When we look at how they see their futures, how they navigate the world today, what hope they have for the future, we see how powerfully children are being affected even in countries that are not facing the worst impact of climate change, yet: 75% worldwide and 73% in the UK think the future is frightening, and over half think that humanity is doomed – 56% worldwide and 51% in the UK.

It is crucial, therefore, when thinking about climate distress with children and young people that we take a global or planetary perspective, encompassing and respecting cultural differences and sensitivity as well as thinking about the immediacy or distance to the threats being faced. It is in the similarities that we see a powerful message about how this is affecting all children and young people. For me this means that an intergenerational frame and child-centred climate crisis lens are essential.

Through qualitative research (Hickman, 2019a) carried out with children globally between 2018 and 2023 using Free Association Narrative Interviewing (Hollway and Jefferson, 2013), key differences could be seen in the way that children and young people locate the threat of the ecological crisis as something in their future or happening now. Children and young people in the UK tended to perceive the threats as happening in the future, in a few years' time; and locate their anxiety around their own choices such as feeling able to have children in the future, even at quite a young age such as this 12-year-old girl:

> *I can't imagine having children of my own – it's just not fair to bring a child into this world with no future for them except hunger and wars over water.*

Or this 10-year-old boy:

*I don't think we should be having more children.*

Or this 18-year-old:

*the only way that I can keep my children safe in the future is by not giving birth to them.*

Fewer children and young people in countries facing current threats from the ecological crisis talked in terms of future generations in the same way; they did hold concern for future generations, but as their fears were focused on the immediacy of their own survival, they were even less able to imagine having children of their own, such as this 18-year-old in Nigeria:

*I'm not sure that I'll live long enough to have children.*

Or were incredulous that anyone they knew would consider having children and were angry with their own parents for having them, such as this 22-year-old in Brazil with parents who were environmental activists:

*I feel betrayed by my parents for having me when they knew what the world would be looking like for my generation, why on earth would I do that to another child, it's insane to even think about it.*

There was more clear anger from the children and young people from countries in the global south who were facing immediate danger, and whilst angry, some were also able to understand that their parents had not personally intended to cause them harm such as this 18-year-old in Brazil:

*I guess they hoped that it would get better, they believed that we would have dealt with it by now.*

Or the 16-year-old in the Maldives:

*my parents just didn't see how bad it was going to get, they still don't believe it now.*

Psychologically and relationally, whilst a few young people were struggling with feelings of anger and blame towards adults generally, the majority did not want to point blame towards individuals but rather saw governments, big business and oil companies carrying the greater share of the responsibility.

An important difference is that some children and young people in Brazil and the Maldives talked about the ecological crisis as a past event that had *already* happened, not to be feared in the future (the latter the perspective of children in the UK). It was framed as something that was *historical* to them and their country even if not in awareness generally.

> *the climate crisis has been going on here for a long time, but it's only now that people seem to be waking up or to care; everyone's been asleep'* or *'my parents failed to deal with this, and their parents didn't take action, so now it's left to us to face, but it's getting too late now, why didn't people act before it got this late?*

## Emotional and Cognitive Impact

There is a growing understanding of the impact on mental health; the distress, confusion and anxiety that follows increased awareness of the climate and biodiversity crisis (Pihkala, 2020; Ogunbode et al., 2022) with concern increasingly centred on how this is affecting children and young people (Lawton, 2019; Hickman, 2019, 2020; Hoggett, 2019).

Recent quantitative research has highlighted the relational aspect of this distress with children and young people suffering severe emotional and mental upset because of climate change, knowing about it, witnessing, experiencing its impact and fearing for their increasingly uncertain futures (Hickman and Marks et al., 2021). They are suffering from trauma and prolonged psychological and physical stress; they have symptoms of depression, grief and anxiety; they feel betrayed, abandoned and dismissed by the people in power and governments whom they expected to protect them (UNICEF, 2021).

Research has also examined the impact of direct exposure to traumatic events (wildfires, floods or extreme heat) as well as indirect adverse experiences observed by witnessing the harm being caused to others such as news reports showing animals and people fleeing wildfires or listening to stories told by survivors of traumatic events (Lawrance et al., 2021; Obradovich et al., 2018).

Whilst research has largely framed eco-anxiety in children as an emotionally healthy and congruent response to environmental crises there remain contrasting and competing theories surrounding its nature and treatment, with eco-anxiety variously framed as pre-traumatic or post-traumatic stress, collective trauma, intergenerational trauma and Adverse Childhood Experiences. Whilst there is a general recognition from professional bodies that eco-anxiety should not be diagnosed or pathologised (RCPsych, 2021), there are also attempts to categorise and differentiate between 'normal' and 'abnormal' forms of eco-distress in children. There is also some debate about how this might be felt differently by children and young people compared to adults. What is certain is that eco-anxiety is an emergent mental health problem that we are identifying amid the unfolding climate and biodiversity crisis itself.

## Therapeutic Approaches, by No Means Definitive Include:

### *Reframing Eco-Anxiety*

Mary Jayne Rust (2021) describes feeling into the disorder of our times and allowing ourselves to be affected by the climate crisis. Sally Weintrobe (2021) talks about the culture of uncare stopping us from feeling sad about the environment and climate crisis.

In my practice I take the approach that children are feeling climate anxiety because they care about the planet; they should be supported to feel proud and glad that they care, rather than have their feelings questioned or attempts made to configure or label them as 'right or wrong', 'good or bad'. This gives permission for the feelings without trying to reduce them; they can be understood, they have meaning, and the fear attached to feelings of anxiety may be reduced through this understanding. I talk about the relationship between eco-anxiety and eco-empathy, eco-compassion, eco-community, eco-understanding, all of which can emerge from reframing eco-anxiety, moving into a different relationship with it rather than trying to manage or get rid of it.

### *Climate Crisis Lens*

In practice we could look through a 'climate crisis lens', similar to a trauma lens which would mean observing a child's attachment, resilience and stress responses. This might help us to stop trying to understand climate distress in children through adult eyes. Adults have mostly grown up without climate distress, learning about it later in life after attachment patterns have been established; by contrast, children grow up with this knowledge from early in life, and feel and process it differently. Developing a secure attachment pattern today in early childhood means incorporating awareness of the insecurity of the planet into this internal frame of reference. A climate crisis lens would help us to develop the empathic understanding that the climate crisis trauma that children are facing today can lead to them organising their mental, physical and emotional lives around patterns of either re-living or avoiding the insecure and traumatic feelings that are being normalised as we face the worsening physical crisis of climate change. Just as the heat increases and ice melts and sea levels rise, our climate distress will continue to develop in parallel.

### *Therapeutic Triad*

Another way to understand how climate anxiety can affect children is to understand that therapeutically we are no longer working in a therapeutic dyad (therapist and client) but a therapeutic triad (therapist, client, and the planet). Children's attachment to the planet is often seen through love of, identification and empathy towards, and relationship with animals and nature. Climate distress in children is often triggered by the indirect trauma of knowing how animals suffer from climate change. They share the innocence of the animals, and so can often express their own distress through their imagining what it is like for animals to be having to face an increasingly hostile and threatening world that they have not created. This can be particularly acute when becoming aware of feelings of unfairness and injustice.

### Mature Defences

There are many good discussions about defences against climate anxiety. I would suggest that we couple these with encouraging the development of mature defences as part of resilience building, including acceptance, altruism, courage, compassion, emotional self-regulation, mindfulness, gratitude, humility, humour, patience, respect, some short-term suppression (time out), tolerance, self-talk and love for the world.

### Climate Anxiety Scale

Understanding the scale of the distress is important. Too often discussions about climate anxiety assume that everyone has similar feelings, on a similar scale. Children have often told me that when they try to talk with adults about their feelings about climate change, whilst the adults seem to understand, they also often do not want to listen to children's distress and try to use reassurance and positive hopeful stories to mitigate the upset. It is understandable that hearing children's distress can be unbearable and invoke feelings of helplessness, but not listening leaves the child feeling misunderstood, as though there was something wrong with their feelings. Children often tell me that whilst they are grateful that their parents helped them get therapy, they also wish that their parents could just try to cope more with their feelings. The model that I have found of some help in this has been the conceptual frame which differentiates between Mild, Medium, Significant and Severe. I have since added Critical as a category to reflect more recent clinical observations from practice with increased numbers of young people talking about self-harm and suicide in parallel with worsening climate breakdown. The table below summarises some key points from the model.

### Dreams, The Imaginal and Personification

Robertson (2021) describes approaching the Anthropocene as a dream citing Thomas Berry's (1988) idea of an earth dream. This is a way to address our self-importance and is a transformational threshold through which we might enter a

*Table 8.1* Psychosocial Climate Anxiety Scale

| | |
|---|---|
| Mild | Feelings of upset (anxiety, distress, fear) are transient and can respond to reassurance.<br>Strong focus on hope and optimism, strong defences against hopelessness.<br>Trust that others will take action. |
| Medium | Upset more frequently, feelings may be stronger and last longer.<br>Reassurance is less effective.<br>More doubt in optimistic solutions.<br>Less trust that others will take action. |
| Significant | Minimal defences against climate anxiety.<br>Much harder to mitigate distress and take comfort from positive stories.<br>More likely to be feeling more complex feelings of guilt and shame.<br>Little faith in others' willingness to take action.<br>Can have a significant impact on relationships with others if they do not share concerns. |
| Severe | More likely to feel intrusive thoughts about the climate crisis.<br>Can be a significant impact on daily life (difficulty sleeping, difficulty resting, spending time with family or friends).<br>Can feel driven by anxiety, urgency, and fear about the future.<br>No belief in other's ability to care. |
| Critical | Huge loss of personal security in the world (nowhere is safe).<br>Nightmares of global and social collapse.<br>Likely to be severe distress possibly including thoughts and/or actions of self-harm and suicide. |

different relationship with climate anxiety and stop seeing dreams as emanating from our egos. Some dreams may be neurotic reflections or wish fulfilments, but others are not ours. They visit us. We are in the dream. 'So, what might be in this earth dream that is dreaming us?' (ibid.: 73). Romanyshyn (2013: 200) also talks about dreams opening us to another way of seeing the world, that 'psyche is weaving a new dream of who we are and how we are in the world'. In practice I hear children dreaming about food shortages, of having to kill their parents or being killed by their parents because of social collapse and breakdown; they tell me dreams of animals speaking with them, asking them to help with the crises, and strange hybrid creatures, half human and half monster, taking over the earth. Children often have a close relationship with the imaginal and talking about how their climate anxiety dreams can help them to 'play with the imaginal landscape of the work' (ibid.: 145).

With dreams we can use active imagination and dialogue with the animals and people who appear with messages. These dreams bring stories about how children are struggling to make sense of a waking world that often makes little sense. Why else would they be living in a world where people in power were saying that they cared whilst so often acting as though they did not. Weintrobe (2021: 5) says we need 'a nightmare and a dream to understand what is wrong and to imagine a better world'. We cannot see the future, but

(a) (b) (c)

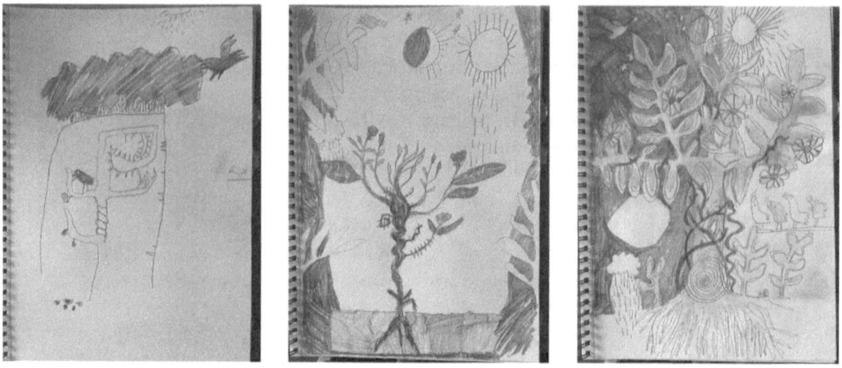

*Plates 8.1* (a-c) Images of 3 Plants created by a child representing: a) Past, b) Present and c) Future.

we can engage with the imaginal which can allow us to feel our way forwards. This is also a way to talk about liminal spaces, in-between old and new worlds. The climate and biodiversity crisis has grown out of the old world, and to face it we need to develop not only an understanding of how we got here, where we are today, but also where we would ideally like to get to in the future. An exercise I use to open this discussion is to invite the child to imagine themselves being a tree, plant or flower, and draw yourself in your life with a past, present and future image. This uses personification (Rowan, 2010) and art to help create a frame to explore rather than classify feelings.

### Stories

There are many books being written to help us talk about climate change with children. One I use as a way of talking about how to engage with the complex thoughts and feelings around this is *We're Going on a Bear Hunt* (Rosen and Oxenbury, 1989). It is often familiar to children (and parents) and helps to understand the need to go through difficulties such as fields of grass, rivers, mud and forests as metaphors for anxiety, fear, feeling stuck or lost rather than avoid, suppress or get rid of distress.

### Resilience

In addition to facing the challenges of feelings and thoughts about climate change, I use the diagram below to have conversations about the benefits of challenges when tackling something as big as climate change or climate anxiety. We all need grit in our compost to help us to grow emotional resilience. Reframing obstacles and challenges as grit can help to discuss emotions in a less binary (good/bad) way and reframe frustrations when climate activism seems to not produce immediate results. Seeds grown in compost without grit do not develop into resilient plants; they grow resilient when their roots are confronted with grit or obstacles that have to be grown around. Plant pot 3 where obstacles are more like rocks is the tougher environment in which to thrive; here I use the metaphor of understanding how personal and planetary trauma can feel overwhelming and perhaps needs the rocks to be broken down into smaller sizes. Community, working with others and compassion for the impact of trauma and seemingly insurmountable tasks can help with this understanding for children.

*Plate 8.2* Illustration of Flowerpots by 'Lisa Made It'.

### *Stop, Breathe, Feel, Think, Act*

In therapy I have found children can be understandably frustrated by exploring feelings about the climate crisis and would rather focus on physical action as a way to reduce their distress. I must agree with this argument:

> *I'd feel ok, well not completely ok, but better anyway if everyone was taking action on the climate crisis. If everyone was doing everything they could do, I'd feel ok. So why would I want to stop feeling distressed when they just need to take action to help my distress!*
>
> (10-year-old – US, 12-year-old – UK, 16-year-old – Mexico, 12-year-old – Maldives)

However, we also need to talk about how to be ok with not being ok whilst acting on the climate crisis (Hickman, 2023). It isn't going to be fixed next week, and living with that level of distress can be hard to cope with. We are increasingly seeing children and young people self-harming, suicidal and killing themselves in response to their distress. This needs a clear and containing response. Whilst not generally a fan of formulaic therapeutic responses, when the feelings are so bad that it is hard to know how to survive, or panic is taking over, I use the following, borrowed and adapted from PADI (Professional Association of Diving Instructors) who teach the Stop, Breathe, Think, Act formula to scuba divers because anything other than this will risk sudden and life-threatening situations underwater. I adapted it for climate anxiety, including *Feel* to honour the need to take both internal action (or activism) alongside external action (and activism).

### *Emotional Biodiversity*

Hollis (1996) argues for the importance of guilt, grief, loss, anger and despair in finding meaning from our experiences, new life and gravitas in what he calls 'dismal places'. Many authors such as Pihkala (2020, 2023) explore the range of emotional responses to climate change and argue for all emotions to be validated and understood. With children and young people, I have used the construct of emotional biodiversity (Hickman, 2021, 2023) to talk about the mixed, competing and contrasting feelings they may be having. This helps avoid the false binary of positive versus negative feelings, making space for hope and hopelessness. Glove puppets help; lions and rabbits often have a lot to say about how they feel!

Climate aware therapeutic models argue that rather than diagnosing and pathologising eco-anxiety in children, we should be supporting them in making sense of their distress, finding meaning in their changing world, developing community and collective approaches for support, challenging climate denial and embracing the range of emotions found through this including grief, despair, depression, radical hope and empathy.

A depth psychology approach (Hollis 1996; Hoggett, 2019; Weintrobe, 2013; Lertzman 2019; Randall, 2019) argues that this very descent, depression, experience of grief and loss gives meaning to the experience of waking up to the climate crisis. Looking under the surface and listening to children's voices requires us to be curious, to use deep listening (Hoggett, 2019), to show respect to all forms of emotional expression (Hickman, 2020; Pihkala, 2020; Weintrobe, 2021), show humility to feelings (Hollis, 1996) and to develop climate aware psychological models that can help children to navigate these unprecedented challenges, both internal and external in the world today.

The worst statistic in the Hickman and Marks (2021) research was that 48% of children and young people (worldwide and in the UK) said that they were dismissed or ignored by other people when they tried to talk about climate change. Therapy with children and young people around climate distress has its dilemmas, but it can, without question, address this injustice, moral injury and relational harm by finding ways to talk with and listen to children without dismissing their concerns or terrifying them.

## Note

1 The clinical examples used are drawn from a) research with children and young people globally which has ethical approval from the University of Bath and all examples used from this have permissions from participants and are fully anonymised (age and gender and country are accurate).

Other clinical examples are composite practice examples rather than individual narratives, reducing the risk of any one individual being recognised. Full permissions were also given by all clients, and they understand that these clinical examples are composite rather than personal. Permissions have been given for drawings to be used anonymously.

## References

Berry, T. (1988) *The Dream of the Earth*. San Francisco, CA: Sierra Club Books.

Hickman, C. (2019) Children and climate change: Exploring children's feelings about climate change using free association narrative interview methodology. In: P. Hoggett (ed.) *Climate Psychology: On Indifference to Disaster*. London, UK: Palgrave Macmillan.

Hickman, C. (2020) We Need to (Find a Way to) Talk About … Eco-Anxiety. *Journal of Social Work Practice*, 34(4): 411–424.

Hickman, C. (2023) Feeling OK with Not Feeling OK: Helping Children and Young People Make Meaning from Their Experiences of Climate Emergency. In: L. Aspey, C. Jackson, and D. Parker (eds) *Holding the Hope: Reviving Psychological and Spiritual Agency in the Face of Climate Change*. Monmouth: PCCS Books, pp. 183–198.

Hickman, C., Marks, L., Pihkala, P., Clayton, S., Lewandowski, R. E., Mayall, E. E., Wray, B. and Mellor, C. (2021) Climate Anxiety in Children and Young People and Their Beliefs About Government Responses to Climate Change: A Global Survey. *Lancet Planetary Health*. www.thelancet.com/action/showPdf?pii=S2542-5196%2821%2900278-3

Hoggett, P. (2019) Getting real. *New Associations*, 29, Autumn.

Hollis, J. (1996) *Swamplands of the Soul: New Life in Dismal Places*. Toronto. Canada Inner City Books.

Hollway, W. and Jefferson, T. (2013) *Doing Qualitative Research Differently: A Psychosocial Approach*. London: Sage.

Lawrance, E. L., Thompson, R., Fontana, G. and Jennings, N. (2021) The Impact of Climate Change on Mental Health and Emotional Wellbeing: Current Evidence and Implications for Policy and Practice. www.imperial.ac.uk/grantham/publications/all-publications/the-impact-of-climate-change-on-mental-health-and-emotional-wellbeing-current-evidence-and-implications-for-policy-and-practice.php [Accessed 30 June 2023].

Lawton, G. (2019) I Have Eco-Anxiety But That's Normal. *New Scientist*, 244(3251): 22.

Lertzman, R. (2019) Climate Psychology: On Indifference to Disaster. In: P. Hoggett (ed.) *Climate Psychology: On Indifference to Disaster*. London, UK: Palgrave Macmillan. pp. 25–39.

Marks, E. and Hickman, C. (2023) Eco-Distress Is Not a Pathology, But it Still Hurts. *Nature Mental Health*, (1): 379–380. 10.1038/s44220-023-00075-3.

McAndrews, J. (2023) Changing the World in One Generation: Raising Children to Grow Resilience Amid Climate and Social Collapse. In: L. Aspey, C. Jackson and D. Parker (eds) *Holding the Hope: Reviving Psychological and Spiritual Agency in the Face of Climate Change*. Monmouth: PCCS Books, pp. 199–212.

Obradovich, N., Migliorini, R., Paulus, M. P., and Rahwan, I. (2018) Empirical evidence of mental health risks posed by climate change. *Proceedings National Academy Science USA 23*, 115(43): 10953–10958. www.pnas.org/cgi/doi/10.1073/pnas.1801528115

Ogunbode, C., Doran, R., Hanss, D., Ojalo, M., Salmelo-Aro, K., van den Broek, K. L., Bhulla, N. and Aquino, S.D. (2022) Climate Anxiety, Wellbeing and Pro-Environmental Action: Correlates of Negative Emotional Responses to Climate Change in 32 Countries. *Journal of Environmental Psychology*, 84: 101887.

Pihkala, P. (2020) Anxiety and the Ecological Crisis: An Analysis of Eco-Anxiety and Climate Anxiety. *Sustainability*, 12(19): 7836.

Randall, R. (2019) Climate Anxiety or Climate Distress? Coping with the Pain of the Climate Emergency. (Blog) https://rorandall.org/2019/10/19/climate-anxiety-or-climate-distress-coping-with-the-pain-of- the-climate-emergency/ [Accessed 16 June 2023].

Robertson, C. (2021) Depth Psychology and Climate Change. In: D. Mathers (ed.) *Depth Psychology and Climate Change: The Green Book*. London: Routledge

Romanyshyn, R. D. (2013) Making a Place for Unconscious Factors in Research. *International Journal of Multiple Research Approaches*, 7(3): 314–329.

Rowan, J. (2010) *Personification: Using the Dialogical Self in Psychotherapy and Counselling*. Routledge. London.

Rust, M. J. (2021) Depth Psychology and Climate Change. In: D. Mathers (ed.) *Depth Psychology and Climate Change: The Green Book*. London: Routledge

Rosen, M. and Oxenbury, H. (1989) *We're Going on a Bear Hunt*. London: Walker Books Ltd.

RCPsych (2021) Information for Parents and Carers www.rcpsych.ac.uk/mental-health/parents-and-young-people/information-for-parents-and-carers/eco-distress-for-parents?searchTerms=climate%20anxiety [Accessed 1 November 2023].
UNICEF (2021) *The Climate Crisis Is a Child Rights Crisis: Introducing the Children's Climate Risk Index*. New York: United Nations Children's Fund.
Weintrobe, S. (2021) *Psychological Roots of the Climate Crisis: Neoliberal Exceptionalism and the Culture of Uncare*. London: Bloomsbury.

Voice 7

# Climate Anxiety has taught me this, so far ...

*T.M. Walshe*

In early autumn of 2021, I was sitting on the hard leather seats of our public mental health clinic. On this particular afternoon, my counsellor was talking about my climate anxiety – something she didn't seem to want to talk about. What I heard helped me understand why so many young people are anxious about the future we're creating, but not in the way I expected. But before I explain exactly what changed my assumptions about society's perception of climate anxiety that afternoon, I want to share some thoughts, experiences, and questions for deliberation.

Today's youth are in an extraordinary situation. From time spent in the climate fight and striking every Friday for the past two years, my friends and I have needed to explain how to read trend graphs to fossil fuel experts with PhDs, talk about why CFCs aren't what is causing climate change to our teachers, and de-escalate fully grown angry adults. No matter how I think about this, I always come to the same conclusion – it is ridiculous. I can never comprehend exactly how we ended up here. However serious or moderately reasonable I may appear, it is very important that people, in order to see me as a real person, understand that I have not, and never will be able to take this on and understand it without it weighing heavily on my life.

I often watch timelapses of $CO_2$ concentration in the atmosphere, or global surface temperature from the 1800s until today. When I watch these, I think about the whisper of time that our youngest generation has been in existence. I watch what happened when the planet started to warm, and then I watch the year 2007 go by. That is the year I was born. One of the most agitating thoughts to cope with as I watch this, is the idea that the future of the Earth I am sentenced to live the rest of my life on was already being written out decades before I got here. I am reminded of the expression, 'your work has been cut out for you'. It feels an awful lot like my life has been cut out for me.

Despite many injustices and anxieties, I am eager to help fix this problem. If people could understand anything at all about this, I wish they understood that. The hardest days are those that are filled with family, friends, teachers, or strangers I respect pushing back, giving up, or ignoring the problem. I'm

DOI: 10.4324/9781003436096-19

ready to work together with all of the people who disagree with me – I wish more than anything that they were willing to work with me.

This brings up a question I would like to understand very much: why is it taking so long for people to recognise the validity of why young people are experiencing climate anxiety? If everybody understood why, and the majority of people wanted to help fix the climate crisis, I can't see why I would be anxious anymore.

Let's return to that afternoon in the fall of 2021. While I was sitting there at my local public mental health clinic, it occurred to me that I did not feel like anybody in that room was trying to understand what it felt like for the hundreds of millions of kids and young people and people in the most affected areas alive today, looking ahead at a climate-crisis filled future, and what we realistically need to combat this grief. As I was thinking this, my counsellor was explaining how I needed to learn to accept climate change and couldn't get my hopes up about taking action. She looked at me and said, 'You are not going to change the world ****. You are not going to change the world'.

When I thought about that day, I used to feel upset. Now, in these days and troubled times of the climate fight, I feel grateful. I feel grateful because my counsellor gave me and thousands of other youth something she will never realise. She gave us something to get up and disprove every day.

T. M. Walshe. Climate Activist, age 15, British Columbia, Canada.

Chapter 9

# Eco-anxiety in the Therapy Room

## Affect, Defences and Implications for Practice

*Trudi Macagnino*

### Introduction

In this chapter I develop a discussion following findings from my doctoral research on how the climate and ecological emergency (CEE) emerges in the therapy room. A fuller account of the method and findings can be found in Macagnino (2022) where I conclude that therapy needs to be re-visioned as a collective eco-psycho-social endeavour rather than a purely individualistic one. Here I elaborate and discuss in more detail what is meant by this revisioning of therapy, and focus on the implications for practice. These include the importance of holding hope, the need for therapists to process their own feelings and to be more proactive in terms of responding to clients' cues. The notion of a traumatised sensibility in relation to the CEE is implied.

### Summary of My Research

I provide this as a basis for discussion about clinical practice.

- A paradox emerged in interviews with therapists and clients. In spite of care and concern about these issues, and value being placed on the natural world as a restorative and spiritual resource, there was a notable absence of feelings relating to the CEE.
- Clients tend to tentatively or even unconsciously bring their concerns to therapy by means of 'side-mentions' which are then side-stepped as small talk by therapists.
- Feelings concerning the CEE were present but often beneath the surface, being inferred from dreams, my own counter-transference and subsequent interpretations. Grief was expressed both directly and indirectly through free association to other losses. Guilt was often expressed directly and was intensified for participants with children and grandchildren. Fear emerged in dreams (wolves and ferocious dogs) and through associations to apocalyptic scenarios. Anger and frustration were directed towards those

DOI: 10.4324/9781003436096-20

seen as unconcerned or thinking they have all the answers. Powerlessness was commonly expressed.

- Unconscious defences used by both clients and therapists included disavowal, splitting and intellectualisation. These mechanisms enabled both clients and therapists to keep the threat out of shared awareness. These are more fully discussed in the presentation of research (Macagnino. 2022).
- There was a tendency for the outer socio-political world to be kept out of therapy with clients and therapists, adhering to a socially constructed view of what therapy is for. Professional and ethical guidelines, and organisational requirements, further cement the collusion and defence. Norgaard (2011) underlines the power in this kind of cultural denial – 'cultural denial is so effective precisely *because* it is invisible and thus feels "natural"' (2011: 133).

## Implications for Therapeutic Practice

The complex interplay of feelings and defences referred to above, which oscillate as therapist and client navigate the terrain of climate emotions, means that clients and therapists do not routinely engage with their feelings about the CEE and ultimately, therefore, do not connect with their agency to act for change.

### *Doing Our Own Work*

As therapists we need to become 'climate aware' by addressing our own feelings about the CEE in order to help clients bear the situation and 'stay with the trouble' (Haraway, 2016). More than a simple cognitive awareness, becoming climate aware is difficult, painful emotional work which involves accepting that which cannot be repaired and facing the prospect of an unsafe, unliveable world. It is likely to feel unbearable at times. Our feelings and defences need to be understood and worked through before we are able to work meaningfully with such material in our practice.

We may fear opening up to distressing feelings. Attempting to maintain our equilibrium we may stick to comfortable topics. Therapist Sarah expressed her fears:

> *I don't want to live my life like that every day … I don't want it to bother me to the point where it starts to take away from my enjoyment of life.*

Therapist Rob was more conscious:

> *It's very, very difficult to engage with the subject of climate change, and destruction of habitats … we're literally falling over the edge of a cliff. I don't*

> *know if we're able to really think about that ... it's very crippling, to even begin to think about it and so I think that we avoid it. The clients are avoiding it. I'm avoiding it. We're all avoiding it.*

When clients make side-mentions, they are signalling a wish to both explore *and* to ignore. The question they are unconsciously asking their therapist is 'Can you bear what I am finding unbearable' (Weintrobe, 2021b). Clients require the safety of knowing their therapist *can* bear it, not pretend the world is safe. They need their therapist to contain them as defences drop away and they descend into grief, terror, rage and remorse.

Kassouf (2022) points out:

> until analysts are able to speak about this on-going trauma, the majority of people with whom we work clinically continue to be left alone with their unarticulated fears about what is happening on the planet. Together, we need to name the dread and try to make it potentially tolerable.
>
> (2022: 73)

Being with the client in this way is not easy. We may feel unable to provide a stable, holding environment; our assumption that the external world is stable and safe no longer holds. We are constantly reminded that our world is unstable, unpredictable, chaotic and threatening. We may find it difficult to speak to clients' pain about the CEE because we have nothing therapeutic to offer and feel doubly powerless. Working with our own feelings about the CEE is different to working through other traumas we have experienced. Rather than reflecting on something from the past, the CEE presents us with trauma in the present and future, we are 'in the thick of trauma' (Kassouf, 2022: 72). Climate trauma is not something to be overcome but to be lived with, requiring a very different approach to the way we are trained.

Processing our own feelings about the CEE in training groups, supervision or with a 'climate buddy' (Allured, 2022: 342), will help us engage with clients. Experiential practices such as grief-tending, climate cafes or *The Work that Reconnects* may also be supportive and useful. The Climate Psychology Alliance has been instrumental in developing resources and supportive spaces for therapists to explore their personal experiences and clinical practice (Tait et al., 2022). This work towards becoming climate aware will help us contain the client, enter a transitional space with them and give a sense of 'we are in this together'. We can explore their feelings from a shared vantage point and ultimately help them find some agency with which to engage with the problem.

## Being More Proactive

A non-directive stance where the client, or their unconscious, leads underlies many therapeutic approaches. However, there are some problems with this stance. First, there is evidence that clients want their therapists to take the lead (Cooper et al., 2019) and are often relieved when asked a direct question. During the research participants were asked directly, 'Can you tell me about your thoughts and feelings about climate change and other environmental problems?' and were willing to explore these. I also asked about their experience of being interviewed:

> *I've really enjoyed it. It's like it's actually made me realise I haven't thought about it enough, … I haven't given it enough of my attention.*
>
> (Sean)

> *the conversation that we had made me think I used to have more passion in me about certain things, … and the things that I could do differently, but I think it's also being prepared to challenge and question and stand up for what I believe in, rather than being more passive.*
>
> (Phil)

> *Do you know I've never talked about it for this long in any one time. It's quite amazing what I'm coming out with.*
>
> (Elaine)

These extracts demonstrate that when a safe space is created through supportive and deep listening – skills therapists have in abundance – proactively exploring feelings about the CEE can be positive, enriching and transformative for clients and can help them to recover their agency.

Cooper (2020) points out that intersubjectivity means we can never be truly non-directive. We are always co-creating our experience with another:

> If they try to direct, it will influence the client in certain ways; but if they try not to direct, it will also influence the client in certain ways.
>
> (online)

By not exploring side-mentions about the CEE we are influencing clients, potentially steering them away from this distressing material, reinforcing the perception that therapy is for personal problems and/or we are not interested in this topic.

Although clients may want their therapist to take the lead, they are unlikely to outwardly criticise them if they do not. This highlights the power differential between therapist and client. Sean blamed himself entirely for not bringing the subject up with his therapist and Deborah asked me not to name her therapist in any publications. This increases the risk of unconscious collusion between client and therapist in disavowal of the CEE and makes it even more important that therapists consider being more proactive in the work.

## Holding Hope

Many people feel powerless when faced with the CEE which can lead to hopelessness and despair. Client Elaine was on the edge of these feelings when she said, '*we've already got to the, you know, the point of no return'*. How can we help clients face the reality without slipping into despair ourselves. 'How do we raise alarm bells without bursting our eardrums?' (Ornstein, 2022: 337).

Sell (2022) suggests that a particular kind of hope is required, a radical 'hope against hope … hope in spite of acknowledging the seeming impossibility or extreme improbability of what is hoped for' (2022: 355). This is not a regressive defensive hope based on wish fulfilment but a productive hope that can lead to change. Therapist Sarah showed an example of regressive hope:

> *I don't buy into that lack of hope. … our trusting and faith and belief in the good of the universe … that the universe is supporting us … if I embody that hope that in some kind of way, it's transferred somehow.*

Conversely, productive hope is where reality is faced and powerlessness is acknowledged. We have to endure our own feelings of powerlessness, in terms of being able to help the client and feeling powerless in the face of the CEE. We have to be willing to enter a transitional space, where something from within (inner) is added to external reality (outer), and creatively allow something new to emerge without any desire for a particular outcome.

It is frighteningly difficult to be with a client in the throes of hopelessness, to share similar feelings and to feel powerless to help. We are likely to feel vulnerable and anxious. We need to guard against defensively grasping to assert control by becoming directive, for instance by prematurely making practical suggestions about how the client can assuage distress. Although this may offer a partial catharsis, it prevents a deeper exploration; the fuller truth of the CEE remains concealed. We may believe that the issue has been dealt with, whereas in fact there has been insufficient emotional engagement to drive change.

## Revisioning Therapy as an Eco-psycho-social Practice

Dominant Western psychology has tended to privilege the individual over social and political contexts, driving our understanding of what therapy is for. However, the CEE is a sociopolitical problem and, therefore, challenges our current frameworks of theory and practice (Bednarek, 2019). My findings support an argument for revisioning therapy as an eco-psycho-social practice (Hollway et al., 2022) such that it is broadened out beyond the individual to consider the social, ecological and political issues affecting us because psychological change is intrinsic to cultural and cultural change on a global scale is what is required to tackle the CEE crisis. As Bodnar asserts, 'The person and the larger social field mutually construct psychological reality. The environment, the earth that sustains life, underlies the social field and the person' (Bodnar, 2008: 485).

When a client comes for therapy, we should aim to help them disentangle the complexity of their distress and understand what belongs to their internal world and what belongs to their external world. As psychoanalyst Nancy Hollander says, we need a 'space in our minds for politics in the clinical setting' where 'concerns of critical social issues should be a legitimate part of the analyst/patient dialogue' (Hollander, 2009: 8). Historically, however, references to political or social material have been interpreted as a displacement of intrapersonal problems, a defence against processing inner pain (Hoggett, 1992). This was illustrated by a participant in the research. An eco-psycho-social perspective reverses the logic and instead sees it as 'a symptom when the patient does not talk about politics' (Saba, 2022: 346).

Rust (2020: 89) argues that 'Our relationships with the more-than-human world are woven into, and inseparable from, relationships with self, family, culture and the earth.' She shows how starting in one place, such as exploring a client's relationship with someone, can lead to exploring the client's relationship with the more-than-human, and vice versa. There is an interconnecting web of relationships that needs to be disentangled for a richer understanding to be gained and potential change to occur.

Kassouf (2022) encourages therapists to shift from their often unconscious 'an environmental orientation' (2002: 62) which excludes the other-than-human from the therapeutic process and instead to listen for moments when it enters sessions. Listening for side-mentions related to the CEE is an example of this kind of listening which could lead to deeper understanding of how clients' feelings and personal histories are linked.

I explored these connections with participants and together we built a shared understanding. For example, Sean's close bond with his dog was associated with his relationship with his grandmother, heightening his empathy for animals and his passion for ethical farming. As a teenager Elaine found sanctuary in the tree branches when her parents were warring, leading her to foster and protect trees.

Kassouf (2022) suggests that clients who have experienced trauma in their lives have a 'traumatised sensibility' which can facilitate useful 'catastrophic thinking' (2022: 63):

> Catastrophic thinking took place with those whose interpersonal trauma had not broken them but *broken them open*, making them aware of their own permeability in generative, liveable ways.
>
> (2022: 64)

> traumatized sensibility can draw strength from vulnerability and dependence rather than erect defenses.
>
> (2022: 71)

Several of my client participants shared details regarding traumatic experiences. Their subsequent ability to discuss their feelings about the CEE with me support Kassouf's concept of a traumatised sensibility. However, a traumatic history may also render the client unable to bear further trauma and this possibility must also be considered. Therapists working in an eco-psycho-social way would link the client's past trauma – their vulnerabilities and resilience – with catastrophic thoughts and feelings about the CEE. They would ask broader questions such as, what is the relationship between this client's presenting issues and what is happening in the world? How does the client's history frame their experience of what is happening in the world? How are my feelings about what is happening in the world affecting how I am meeting this client today? Such questions will help judge whether the client has the resilience to engage in catastrophic thinking about the CEE or whether it may be too overwhelming.

Working in an eco-psycho-social way is currently not the norm for most therapists. New elements need to be introduced or existing elements combined in new ways. Some argue for therapy to include a more collective and publicly accessible experience. In her chapter Jo Hamilton describes Emotionally Reflexive Methodologies – group- and community-based approaches to working emotionally with the CEE.

The developing practice of eco-therapy is another form of recrafting involving the inclusion of the more-than-human world within the therapeutic process. It assumes a reciprocity between the human mind and the more-than-human world which plays a significant part in our psychological development and wellbeing (Roszak, 1995). In eco-therapy the natural world can be viewed as co-therapist (Berger, 2006), a secure base (Jordan, 2015), a transitional space or object (Jordan, 2015), or as the therapeutic third vantage point (Berger, 2006). There is a risk that eco-therapy simply becomes a means of seeking solace in nature rather than a true re-integration of inner and outer. However, enabling clients to locate themselves within the more-than-human and recognise our inter-connectedness is arguably critical at this time.

An eco-psycho-social practice could also see therapists becoming more socially and politically active as expressed by this activist/scholar:

> Psychoanalysts, we need your help. ... step off the sidelines. Take your couch out of the office and into the march! Counsel us as you walk beside us. Then offer your unique talents. Shape with us a container for grief, fear, and despair. ... Help us transform these feelings into fiery action.
>
> (Ornstein, 2022: 338)

Action can help us bear the overwhelming feelings evoked by the CEE. Li et al. (2022) argue that mental health professionals have a duty to lead systemic change. They suggest shifting focus from the individual to the social, developing new models of working, educating ourselves and using our professional status to advocate for change in our communities. Alongside this, the importance of emotional work is becoming recognised within activist movements and reflexive activism is developing (Hoggett, 2009; Lawson, 2021) which therapists could support. It is important that this inner work is not just a means of coping with the demands of activism but a true attempt to integrate inner and outer worlds.

Simply adding an ecological lens to therapy or a therapeutic lens to activism does not dispense with the duality of inner and outer, which can be seen to reflect the dualism of society and nature. There is a potential meeting place, perhaps outside of both traditional therapy and activism, where inner and outer come together in a truly eco-psycho-social space.

## Conclusion

Therapists have an important role to play in these times of the CEE but will need to work more eco-psycho-socially. Therapy can support people to live with the reality of the CEE and the difficult feelings it evokes by providing the containment and safety to process them. It can act as a bridge supporting people to connect with their agency and to act for change, countering feelings of powerlessness and hopelessness. However, therapists need to personally engage with the CEE, to process their own feelings and manage the individual and social defences that are a barrier to this. As the reality of the CEE continues to make itself felt and we emerge from our 'climate bubble' (Weintrobe, 2021b) we are likely to experience shock, feel vulnerable, angry, traumatised, ashamed, afraid and more. This collective trauma will be the context in which we are working; we will need the support of our colleagues, supervisors, training organisations and professional bodies to adapt in order to navigate these waters.

## References

Allured, E. (2022) Dreaming Beyond the Traditional Frame: Climate Psychoanalysis Meets the Press. *Psychoanalytic Dialogues*, 32(4): 341–343. 10.1080/10481885.2022.2090808.

Bednarek, S. (2019) 'This Is an Emergency' – Proposals for a Collective Response to the Climate Crisis. *British Gestalt Journal*, 28(2): 4–13.

Berger, R. (2006) Beyond Words: Nature-therapy in Action. *Journal of Critical Psychology, Counseling and Psychotherapy*, 6(4): 195–199.

Bodnar, S. (2008) Wasted and Bombed: Clinical Enactments of a Changing Relationship to the Earth. *Psychoanalytic Dialogues*, 18(4): 484–512.

Cooper, M. (2020) Non-Directivity: Some Critical Reflections. *Research Findings*, 1 July. Available at: https://mick-cooper.squarespace.com/new-blog/category/Research+Findings [Accessed 15 August 2022].

Cooper, M., Norcross, J. C., Raymond-Barker, B. et al. (2019) Psychotherapy Preferences of Laypersons and Mental Health Professionals: Whose Therapy Is It? *Psychotherapy*, 56(2): 205–216.

Haraway, D. J. (2016) *Staying with the Trouble-making Kin in the Chthulucene*. Durham and London: Duke University Press.

Hoggett, P. (2009) *Politics, Identity and Emotion*. Boulder, CO. Paradigm Publishers.

Hollander, N. C. (2009) When Not Knowing Allies with Destructiveness: Global Warning and Psychoanalytic Ethical Non-Neutrality. *International Journal of Applied Psychoanalytic Studies*, 6(1): 1–11. Available at: 10.1002/aps.183.

Hollway, W., Hogget, P., Robertson, C. et al.(2022) Introduction: A Matter of Life and Death. In: W. Holloway, P. Hoggett, C. Robertson et al. (eds) *Climate Psychology: A Matter of Life and Death*. Bicester: Phoenix Publishing House, pp. 1–13.

Jordan, M. (2015) *Nature and Therapy: Understanding Counselling and Psychotherapy in Outdoor Spaces*. First edition. East Sussex: Routledge.

Kassouf, S. (2022) Thinking Catastrophic Thoughts: A Traumatized Sensibility on a Hotter Planet. *The American Journal of Psychoanalysis*, 82: 60–79. Available at: 10.1057/s11231-022-09340-3.

Lawson, A. (2021) *The Entangled Activist*. London: Perspectiva Press.

Li, C., Lawrance, E. L., Morgan, G. et al. (2022) The Role of Mental Health Professionals in the Climate Crisis: An Urgent Call to Action. *International Review of Psychiatry*, 34(5): 563–570. Available at: 10.1080/09540261.2022.2097005.

Macagnino, T. (2022) Why Aren't We Talking About Climate Change? – Defences in the therapy room. *British Gestalt Journal*, 31(2): 26–33.

Norgaard, K. M. (2011) *Living in Denial: Climate Change, Emotions, and Everyday Life*. Massachusetts: MIT Press.

Ornstein, J. (2022) Psychoanalysts: The Climate Movement Wants YOU. *Psychoanalytic Dialogues*, 32(4): 337–338. Available at: 10.1080/10481885.2022.2090803.

Roszak, T. (1995) Where Psyche Meets Gaia. In: T. Roszak, M. Gomes and A. Kanner (eds) *Ecopsychology: Restoring the Earth, Healing the Mind*. San Francisco: Sierra Club Books, pp. 1–17.

Rust, M-J. (2020) *Towards and Ecopsychotherapy*. London: Confer Books.

Saba, M. (2022) Climate Crisis and Psychoanalytic Responsibility: 'There Is No Such Thing … '. *Psychoanalytic Dialogues*, 32(4): 346–347. Available at: 10.1080/10481885.2022.2090811.

Sell, C. (2022) Medio-Passive Agency in Psychoanalysis: Responding to Hopelessness and Despair in the Therapeutic Relationship. *Psychoanalytic Dialogues*, 32(4): 353–368. Available at: 10.1080/10481885.2022.2082251.

Tait, A., O'Gorman, J., Nestor, R. et al. (2022) Understanding and Responding to the Climate and Ecological Emergency: The Role of the Psychotherapist. *British Journal of Psychotherapy*, 38(4): 770–779. Available at: 10.1111/bjp.12776.

Weintrobe, S. (2021a) *Psychological Roots of the Climate Crisis: Neoliberal Exceptionalism and the Culture of Uncare*. New York: Bloomsbury.

Weintrobe, S. (2021b) On Becoming Climate Aware as a Psychotherapist. *Royal College of Psychiatrists: Faculty of Medical Psychotherapy Annual Conference Survival and Development; Exploring our Internal and External Landscapes*, Oxford, 22 April. Available at: www.sallyweintrobe.com/april-22-2021/ [Accessed 21 July 2022].

Chapter 10

# Climate Silence in the Consulting Room: Waiting for Help to Come

*Paula Conway*

At the beginning of Yoko Ogawa's novel, *The Memory Police*, the narrator recalls a conversation between herself and her mother, when she was a child. Her mother tells her about a time on the island where they live when there were many things that have now disappeared – transparent, fragrant, fluttery things – marvellous things that no longer have names, because on this island objects disappear, or are disappeared, and with them, the memory of those objects. She tells her daughter that one day soon she will have an experience of a disappearance and reassures the frightened child that it won't hurt or even feel sad. She will simply wake up one day, and: 'Lying still, eyes closed, ears pricked, trying to sense the flow of the morning air, you'll feel that something has changed from the night before, and you'll know that you've lost something, that something has been disappeared from the island' (Ogawa, 2019: 3). Reading this novel, I was struck by the passivity in the text, the bland acceptance of these losses. Wonderful things are 'disappeared' from the island, but the inhabitants barely feel they have gone. Soon they won't remember them. There's nothing to worry about, it doesn't hurt.

Why do my patients rarely, if ever, bring climate change and the environmental crisis to their therapy? This question had been puzzling me for some time, but one morning in November last year, as I sat looking across my consulting room out of the window at the cherry tree, the first to lose its leaves, looking bleak and bare, as if it had died, my mind drifted back to the summer. I recalled that when the tree was resplendent with auburn leaves and dark red cherries, there had been, unusually, no bullfinches darting in and out of the branches. I noted it yet didn't, asking my husband once or twice, have you seen any bullfinches this year? But otherwise getting on, doing what needs to be done, not letting myself dwell on the disappearance of those comical-looking creatures, always in pairs, their pink satin waistcoats stretched tight across bulging bellies. I know that bullfinches are in decline, but not here, surely, I thought that morning, where they have always triumphantly plucked the last cherry from the tree. They will be back next year, I told myself.

Then, hearing those empty reassurances, I wondered if I was experiencing the passivity towards a 'disappearance' portrayed in Ogawa's novel, or the

DOI: 10.4324/9781003436096-21

pervasive apathy towards the environmental crises that Searles proposed as far back as 1972. A form of *don't worry, it doesn't hurt,* or *they'll be back next year,* that prevents us from really thinking about the escalating environmental catastrophe. A defence against mourning – 'you won't even be particularly sad', says the narrator's mother (ibid.).

When I reflected carefully on that morning's work, I had another thought, different from apathy or a failure to face reality and mourn. I wondered if perhaps there was something I wasn't hearing in myself and in my patients. Why did my mind take me, right at that moment, to the cherry tree, the summer, the bullfinches or, more accurately, the disappearance of them? Were my patients trying to bring me their anxieties about a loss that is so overwhelming it cannot be mourned, or something so broken they fear it cannot be repaired? I was reminded of Henry Rey's (1988: 457) paper, where he writes: 'how very frequently, if not always, help is asked with regard to improving oneself, whilst the real request is how to bring about the reparation of important damaged inner objects without which reparation the subject's self cannot function normally and happily'. Rey describes how patients need to keep alive, through identification, a dying but not yet dead object, because 'so long as they were kept alive there was a chance to repair them … [and] there was hope somebody would come who would know how and would help do it' (ibid.: 467).

Just that morning, I had heard about a mother-child relationship, deeply damaged when my patient was just two, and glued uncomfortably back together ever since. I heard about a broken marriage and a patient, alone for the first time in her adult life, telling me: 'I just want to be able to say sorry for the part I played'. I listened to the painful story of a foster carer's disturbed sibling seemingly reincarnated in the carer's troubled foster child. How can we get better, they all seemed to be asking, when there is so much damage, so much beyond repair? And, as one patient said, 'I know it's grim right now, but I must have some hope'. I wondered if my patients were in their own way also telling me about a world that is damaged and broken, a dying but not yet dead object, and how we are all in some ways responsible but also in other ways not, and about a need for hope that help will come.

There are many theories to explain our collective silence on the climate emergency, in and outside the consulting room, which I cannot do justice to in this short paper. George Marshall (2014) writes that we are not wired to think about a problem that we can't obviously see, that is predicted to harm us in the future and about which we feel individually helpless to address. Searles (1972) hypothesised a whole host of psychological reasons, including unworked-through oedipal rivalry with our children, leaving us unconcerned about them being polluted into extinction. Or, rather than facing and mourning the loss of our childhood world, we omnipotently spoil it. As in the Memory Police, where eventually even parts of peoples' bodies go missing and even these are apathetically adjusted to, Searles writes: 'we shall have essentially nothing to lose in our eventual dying' (ibid.: 367).

In *The Psychological Roots of the Climate Crisis*, Sally Weintrobe (2021) writes compellingly about the multiple forms of denial and disavowal we use to not know about our role in climate change, embedded in a neo-liberal capitalist culture of consumerism and 'uncare'. Brenner (2019) describes how those who are concerned about climate breakdown can be scorned as if they are the hysterical chicken in the fable Chicken Little, who panics when an acorn falls on her head and believes the sky is falling in. Brenner (2019) also talks about our pre-oedipal maternal transference to the environment, believing we can make a mess and rely on 'mother nature' to clean it up, like our mothers cleaned up our baby mess, literally and emotionally. Essentially, though, there is agreement that the causes of our collective silence, denial and inaction are multiply determined and clearly not confined to the consulting room.

But silence in the consulting room does, of course, raise questions of technique. What to do, when, overtly, patients (my patients, at least) do not bring concerns about the environment to the sessions? Totton (2014: 213) wonders how much as therapists we should be interpreting dreams and associations in terms of the climate crises. That we must now be thinking *ecosystemically*, which 'contradicts treating the client as an isolated monad whose experiences and problems relate purely or primarily to their internal reality'.

Do I need to be more proactive, and listen more closely with an ear particularly alert for unconscious communications that may be conveying overwhelming and annihilating anxieties about climate change and the many disappearances that are happening all around us? And if so, what is the most helpful way to respond? In this chapter I describe some of my own attempts to think *ecosystemically* in my work, with two examples, one where I did, albeit clumsily, make an intervention, and one where, so far, I have not.

## A Trip to Goa[1]

Yasmin tells me she is going on holiday to Goa to practise yoga, meditate and be away from all material things for two weeks. She is excited but terrified of flying and always drinks alcohol and takes Valium before any flight. We discuss how this connects to having a mother who was mentally ill when Yasmin was a baby, and her difficulty in placing her life in another's hands.

Then she talks about an Indian film she saw, based on a true story, in which a boy became separated from his mother by jumping onto a train when he was about five. The train leaves the station and arrives in another city. The boy is eventually adopted and doesn't see his mother again for over 20 years. 'He doesn't even remember he had another mother,' Yasmin says. 'I know I have a mother, but somehow, I identify with that five-year-old child. I feel I lost my mother somewhere along the way.' She has tears in her eyes. Later in the session she describes the long walks she takes in the woods with her dog, and how being connected to nature makes her feel less depressed. I comment that she seems passionate about nature yet hasn't mentioned any mixed feelings about taking a long-haul flight to Goa. She looks at me in hurt puzzlement. 'I don't understand,' she says, 'what one has to do with the other?' I immediately feel, simultaneously, *what a clumsy interpretation!* but also *am I going mad – isn't it obvious?* I continue, saying, 'Well, you know, in terms of the impact of flying on the environment, which you seem to love.' She pauses, then says: 'I do care very much about the environment, and Goa has a terrible record of cutting down forest which cuts off the water supply to rural villages. I just feel helpless to make a difference – whether I go or not, the trees will keep coming down.' Nothing more was said in that session about her forthcoming holiday or my rather crass interpretation.

In the following session, Yasmin says she has been thinking about my comment on the environment and she still doesn't understand why I said it. I say that perhaps she found it jarring because it was so far from what was on her mind, which was about feeling cut off from her mother, and my comment made her feel cut off from me. Like the little boy in the film losing his mother, in that moment, she felt she had lost her therapist. She agrees, and again I am reminded of Henry Rey's 'damaged objects'. Yasmin has very directly told me she believes she is in analysis to repair both her parents – her mother who had two psychotic breakdowns during Yasmin's childhood and her father who is, as she puts it, 'a middle-class alcoholic'. When I tried to link two things that were not connected in her mind, I became the mother whose mind is elsewhere, with another agenda that makes no sense to Yasmin. But she is also her mother in that moment, breaking the link between flying and climate change, and, for a psychotic moment, leaving me as the incomprehensible Yasmin thinking – *but isn't it obvious?*

Interestingly, Yasmin did bring up the issue of deforestation, about which it is true there is nothing she personally can do, but didn't want to talk about the flight, which is within her power to address. I think my comment on the flight was experienced as me asking her to feel guilt, and as she is already overwhelmed with guilt about how much hate, despair and dependence she feels towards her mother, this was a step too far. To take responsibility, or even to feel some ambivalence about the flight, would mean to give up her sense of helplessness, just when it is what she most wants to communicate to me. For me to challenge her at that moment was to miss the point and enact in the transference an out-of-touch parent with her own incomprehensible agenda. However, a few sessions later, in tears, Yasmin tells me again how worried she is about her mother, and, *using exactly the same words,* says how helpless she feels to make a difference. At this point I can link her helplessness back to our conversation about the environment – that she is concerned about her mother and Mother Earth but feels helpless to do anything. Putting it this way, without evoking any guilt or, at this stage anyway, any need for her to do something about it, she agrees and is not persecuted by the link.

This reminds me of Mary-Jane Rust (2008) linking the content in a patient's dream with wider environmental issues, and how she too is rebuffed by the patient who is angry at her bringing in her own green agenda. Rust writes that with hindsight she could have put her interpretation in a way that may have been more palatable to the patient, and wonders whether her patient felt accused of an environmental crime. I am sure that I too could have talked to Yasmin in a less judgemental way, and I think I did accuse her of an environmental crime, consciously linked to my anxiety about the environment, but perhaps also because of surprise and annoyance at her announcing an unexpected two-week break, out of sync with my breaks, and some envy of her retreat to Goa!

## The Dream

Sarah has been attending weekly therapy for about six months when she brings her first dream to a session – a big dream that seems to illuminate internal and external anxieties with images that feel both personal and collective.

In the dream, Sarah is walking on the beach when suddenly a figure appears in front of her, a small, ragged child wearing a peaked cap, a bit like the Artful Dodger in Oliver Twist. The child beckons her to follow him and leads her away from the beach into a house full of discarded, useless objects. Parts of computers, broken furniture, empty plastic bottles piled in corners. In the house she finds her mother collapsed on a chair. She picks her mother up and springs in the chair bounce up into Sarah's face. The room feels terribly cold – there are icicles dangling on the inside of the window frames. She carries her mother to another room, but this room is so hot and humid it is almost impossible to breathe. Climbing over the debris, she carries her mother from room to room, but none of them is a comfortable temperature. Then she realises the house is moving and when she looks out of a window, she sees it is drifting back towards the sea. The child beckons her again, this time to an upstairs room, and with her mother still in her arms she climbs out of the window and onto the roof. The child climbs out with her, only now it is not a child but a black cat. Sarah says she can't remember if a black cat crossing your path is meant to be good or bad luck. She realises that her mother has died and as she holds her dead mother in her arms, her mother turns into a large bird like a pelican or a heron and flies away. The cat stalks away with its tail in the air. Sarah wakes with tears on her face.

Sarah has many associations to the dream that relate to her childhood, current personal circumstances and her ambivalence about having therapy. She is still mourning for her mother who died a year earlier. She often walks on the beach and has fantasies of taking off her clothes, wading into the sea and never coming back. She feels the child in the dream is possibly me, artfully leading her back from death into this uncomfortable room that is meant to be helpful but doesn't always feel it. It is too late, she says, she should have had therapy years ago, when she was a child. Even though her mother has died, Sarah feels she will never be free of her mother, whom she carries her around in her arms, feeling weighed down. Many things feel broken and damaged in Sarah's internal world and in her past and present relationships. She feels both aggrieved but also desperate to make things better, for some hope.

Of course, I am also having associations to this dream. I think of the rooms, either too hot or too cold, as the effects of climate change. Her collapsed mother is Mother Earth collapsing under the weight of our debris and the amount of stuff that we accrue but don't need. The house floating and Sarah climbing onto the roof reminds me of the images we see so often now of floods around the world and desperate people taking refuge on their roofs. Sarah's despair that it is too late resonates with my own despair about the ransacking of our planet and the unbearable consequences.

Sarah makes none of these associations and says nothing that I could easily link to my thoughts. It is the first dream she has brought to her treatment. Fragile and desperate to feel better, she is facing layers of intergenerational trauma and grief, and frequently feels suicidal. Do I wade in and make a link to the wider problems facing the planet? I am holding my breath as Sarah talks about the dream and her thoughts. I feel, as Bernstein writes: 'The challenge in this instance is not to interpret at all – certainly not in the moment – to hold an experience that can feel between language, that can leave one with the tension of holding one's rational breath for far longer than any of us can imagine doing' (Bernstein, 2005, cited in Totton, 2021: 29). Although I believe the dream is also an alert to the wider environmental crisis, I feel I must be with Sarah, where she is right now, and tread lightly on this dream. As a supervisor once said to me early on in my career, our task is to bring to the surface that which is blinking up at us just below the water. To cast my line too deep would result, I think, in pulling up a single welly, of no use to anyone. And I have no doubt we will return to the dream again and again during her therapy.

If the natural world is a significant object relationship in human development (Bodnar, 2008), then, as with all important object relationships, we will have a highly ambivalent relationship towards it, precisely because of our dependence on it. The planet doesn't need us. Yet, because of our big heads and bipedality, we give birth to human infants in a premature state needing constant care for many years (Conway and Ginkell, 2014). We are dependent on our primary caregivers for our lives, and this instils in us a terror and hatred of dependence on a maternal object that we cannot control. As well as our love of and need for nature, this dependence on an uncontrollable object makes us also hate Mother Earth – the wild flora and fauna that has the same indifference towards us that Yasmin demonstrates in relation to her long-haul flight. Lertzman (2013: 129) suggests that we replace apathy towards the environment with more complex ideas of 'paradox, contradiction, grief, shock, loss and other affective processes ...' (ibid.:129). I would add hate, fear, rage and guilt to those affective processes, as well as love and reverence.

We need a certain degree of psychic integration to bear guilt and even greater emotional development to experience concern and accept responsibility (Winnicott, 1963). But, as with the oscillation Klein describes between the paranoid-schizoid and depressive positions, concern isn't a solid state that one arrives at; it is sometimes available, sometimes not. Winnicott suggests we need to feel we can be constructive before being able to own our destructiveness. I find this a compelling concept vis-à-vis our agency in relation both to climate change and our work as psychotherapists. When the mother can take the full brunt of her baby's rage and also be there for the baby to make reparation, then the capacity for concern can emerge. I think it is our job as psychotherapists to withstand this rage, recognise reparative moments in our work and thus pave the way for concern that our patients will use in whatever way feels right for them.

I frequently hear my patients asking if there is hope or how to have hope in the face of at times overwhelming despair. I also need to keep alive a sense of hope in the face of what can feel too little, too late. Cal Flyn (2021: 74) explores land that has been abandoned due to war, disaster or political and economic upheaval, and describes the phenomenal scale of reforestation and regeneration in these spaces, of both plant and animal species. She writes: 'All that the Earth may need to soak up enormous, climate altering quantities of carbon is to be left alone'. I began this chapter wondering if I wasn't hearing or interpreting actively enough my patients' unspoken anxieties about the climate crises. I now feel that getting out of the way, bearing the unbearable, and as Segal (1988: 56) wrote about nuclear threat, '[in] dealing with the patient's basic defences, the relevant material will appear because, in fact, below the surface, patients are anxious, even terrified'.

By the time this chapter is in print, the cherry tree in my garden will be in full bloom and laden with purple cherries. Will the cheeky bullfinches be darting in and out of the branches, or have they gone forever? Another disappearance from the island? Having now written about them, I will no longer be cut off from my grief and will mourn them and miss them. I won't forget. I will also hold onto the hope that in some future, near or distant, they may be back.

## Note

1 Client material throughout is amalgamated and details changed to preserve anonymity.

## References

Bernstein, J. (2005) Living in the borderland. In: N. Totton (ed.) *Wild therapy*. Monmouth: PCCS Books Ltd, p. 29.

Bodnar, S. (2008) Wasted and Bombed: Climate Enactments of a Changing Relationship to the Earth. *Psychoanalytic Dialogues*, 18(4): 484–512.

Brenner, I. (2019) Climate Change and the Human Factor: Why Does Not Everyone Realise What Is Happening? *International Journal of Applied Psychoanalytic Studies*, 16(2): 137–143.

Conway, P. and Ginkell, A. (2014) Engaging with Psychosis: A Psychodynamic Developmental Approach to Social Dysfunction and Withdrawal in Psychosis. *Psychosis*, 6(4): 313–326.

Flyn, C. (2021) *Islands of Abandonment*. London: Harper Collins.

Lertzman, R. (2013) The Myth of Apathy. Psychoanalytic Explorations of Environmental Subjectivity. In: S. Weintrobe (ed.) *Engaging with Climate Change. Psychoanalytic and Interdisciplinary Perspectives*. New York: Routledge.

Marshall, G. (2014) *Don't Even Think About It: Why Our Brains Are Wired to Ignore Climate Change*. London and New York: Bloomsbury Books.

Ogawa, Y. (2019) *The Memory Police*. London: Penguin.

Rey, H. J. (1998) That Which Patients Bring to Analysis. *International Journal of Psychoanalysis*, 690: 457–470.

Rust, M-J. (2008) Nature Hunger: Eating Problems and Consuming the Earth. *Counselling Psychology Review*, 23(2): 70–78.

Searles, H. (1972). Unconscious Processes in Relation to the Environmental Crisis. *Psychoanalytic Review*, 59: 361–374.

Segal, H. (1988) Silence in the Real Crime. In: H. B. Levine, D. Jacobs and L. J. Rubin (eds) *Psychoanalysis and the Nuclear Threat: Clinical and Theoretical Studies*. California: Analytic Press.

Totton, N. (2021) *Wild Therapy (second edition)*. Monmouth: PCCS Books Ltd.

Weintrobe, S. (2021) *Psychological Roots of the Climate Crisis*. New York: Bloomsbury.

Winnicott, D. W. (1963) The Development of the Capacity for Concern. In: D. W. Winnicott (ed.) *The Maturational Process and the Facilitating Environment*. London: Karnac.

Voice 8

# Activist journey

*Elouise M. Mayall*

In spring 2020, several things went wrong in quick succession. First was my mother's stage 3 breast cancer diagnosis, which was followed two weeks later by the national Covid lockdown. Since my father is a doctor who manages intensive care clinics, our family considered the not too unlikely risk of losing both my parents. Wills were updated, tough conversations were had, and I as the eldest daughter was emotionally preparing for the task of keeping everyone else afloat. Adding in the violence and aggression that black communities experienced during the Black Lives Matter movement, all whilst trying to complete my undergraduate degree, it is fair to say that I was not doing well.

I did not feel as if I could get the support I needed from those close to me, nor did I want to be a bother with everything else going on, so I turned towards counselling for the first time. Overall, I found it to be a very positive experience and the counsellor provided a lot of support for the aforementioned issues. I really liked her and the grounding presence she provided in a very unstable time and began to entrust her more with deeper concerns.

After a couple of months, I mentioned my climate anxiety, my feelings of guilt, the panicky flutter in my chest, and my frustration over the lack of urgency I saw from those who were supposed to care. Whilst talking I noticed her brow had begun to furrow, head cocked slightly to one side with a quizzical look on her face. She gently asked why I was feeling these emotions so intensely and I tried to explain, but that quizzical look did not go away. I started to feel embarrassed, maybe even a little ashamed, so I diverted the conversation, as this was not another space where I wanted to feel dismissed in a time when I was already feeling very isolated.

We continued having sessions for several months, where I greatly appreciated her support, however I never brought up my feelings about the climate crisis again. I was worried I would need to justify its relevance and learned that this was not the place to talk about it. Thankfully the primary reason I sought out counselling was not for my climate anxiety; however, I know it will forever be an undercurrent in my daily life, interacting with all

DOI: 10.4324/9781003436096-22

the other more 'typical' trials and tribulations that might lead someone to seeking mental health support.

In the absence of professional help, I have relied on climate activism groups for support for my climate anxiety. Having access to a community where I didn't need to justify my concerns made me feel more secure. We do not have the answers, nor the tools to cope with those in crisis, but sometimes the simple validation of the legitimacy of your experience is enough to temporarily quell the anxiety.

I have also found comfort in my fieldwork, where the quiet of the forest and the tasks at hand kept me in the present. Green spaces have long been known to be calming and restorative places and I believe I would struggle in a job where access to these places was limited. Unfortunately, with each year, the effects of the climate crisis are more keenly felt, and I worry I will no longer be able to find the same sanctuary in nature. I count nests failing due to dehydration, as once vibrant mossy forest floors crumble underfoot. As I witness the desiccation of these sylvan cathedrals, I fear that I am running out of places to turn.

Space for climate anxiety must be made in health care settings. With each new COP or IPCC report the evidence screaming how dire our situation is grows louder, but it feels as though only a few can hear it. As the climate crisis continues, the mental and emotional stress that children and young people face will only grow with it. For some, the experience might be mild, for others, very severe. It can also be highly turbulent in its intensity and focus. Whilst the solutions to the climate crisis might be complex, I do not believe accessing mental health support should be. The necessary skills are already there, but practitioners must take time to listen to those calling out for help, without dismissing or minimising their concerns. Everyone has a role in adapting to and overcoming the climate crisis, it is time for mental health practitioners to take up theirs.

Elouise M. Mayall, UK

*Plate 8.1* Elouise shows us just what it means to her to follow her love for the wild forests and creatures within with this photograph of a baby owl. It sits in her hands, being measured, and cared for. Staring at the camera, feathers fluffy and half formed, as if to say 'can you see me here?' There is no human person seen in the photograph, they are an adjunct to the owl. Young people need this care in the face of the climate crisis as well as this owl.

Chapter 11

# 'Climate Mania'

*Garret Barnwell*

I woke one Monday morning to messages on my phone and social media, followed by several calls to my practice. Lawrence Clarke (pseudonym) was an investment banker in a well-known international firm. 'Lawrence hasn't been sleeping for several nights,' his sister concerningly relayed to me over a call. 'I can't describe it any other way other than climate mania,' she said. Clarke also identified with 'climate mania' seeing it as accurately describing his experience.[1] When speaking with Clarke, his speech was pressured, and he jumped from word to word making loose associations about the climate crisis and how to save the world. 'Lawrence, you can come to my practice this morning,' I responded, knowing there was urgency in everything he said. It was clear to me that he was confronting a terrifying truth about the global catastrophe of climate change that is set to make the world, as we know it, unliveable (IPCC, 2022).

In our first session, Clarke cried with fear, joy and pain. He is a loving, warm, easy-going person, and this was magnified in his emotional state. He arrived at my practice carrying a book filled with writing, pictures and messages from God that he had noted down. Words spilled across the pages and his more recent writings were fragmented. He anxiously spoke about the urgency to act, swiftly moving from idea to idea. What was clear is that he was captivated by a calling to save humanity from itself.

In this chapter, I describe the therapeutic journey with Clarke. To my knowledge, this is the first case study to describe a manic-depressive response to the climate crisis. It is my hope that in documenting this, other therapists may benefit, as well as from the Lacanian conceptualisations that informed my approach.

I am a clinical psychologist in private practice, offering psychoanalytically oriented therapy. I work with people recovering from psychosis, drawing on my clinical knowledge and personal experience with family members who have faced such states. Through this work, I have been fortunate to encounter the writing of Lacanian psychoanalysts Apollon, Bergeron, and Cantin (2002; Librett, 2020), Fimiani (2021), Hook (2021; 2022) and Leader (2013), who have largely informed this case conceptualisation.

DOI: 10.4324/9781003436096-23

I begin this chapter with Willy Apollon and colleagues at GIFRIC (*Groupe Interdisciplinaire Freudian de Recherches et d'Interventions Cliniques et Culturelles*), best known for their work in the Clinic of Psychosis, who offer thinking about psychosis and climate change. Librett (2020) has summarised Apollon and colleagues' views as follows.

Apollon and colleagues see the clinic as being located within this historical time where globalisation – and, I would be even more specific here, capitalist expansion – is reconstituting the world as we know it. They attest that forces impressing on the psychoanalytic subject today include: the climate catastrophe; wars; forced displacement; the erosion and reconfiguration of national and group identities; the fortification of borders in response to encounters with displaced persons; shifting social dynamics, cultures and values due to increased urbanisation and migration; information wars resulting from knowledges' democratisation and falsification; and, positively, the increasing empowerment and education of women. Consequently, the present moment confronts us with an ethical choice about responding to these changes. Those living with psychosis often take a strong ethical position towards such societal issues compared to, for instance, people with neurotic structures who rely primarily on repression as a psychological defence (Meyer, 2023). People with psychosis may have difficulties separating themselves from the more-than-human world, and the destruction of the planet in particular (Dodds, 2011).

As with so many people worldwide, Clarke said he had largely been oblivious to the fact that life was rapidly declining. Until recently, he was relatively happy with the comfort and stability that an upper-middle-class existence afforded him. Nevertheless, he mentioned experiencing a subtle underlying disquiet about his lifestyle's wastefulness and the impact that it was having on the planet. Over time, I learned that his worldview began shifting dramatically in the year before we met, after he had a near-death experience due to a severe medical issue. He explained that he grew aware of life's fragility and a void of meaninglessness impressed upon him. For almost a year, feelings brewed under the surface. It was only after he watched the popular movie *Don't Look Up* that his world of being a well-to-do banker, which had once given him clear coordinates, lost its solidity.

As Lacan so aptly reiterates throughout his work, we come into being through the language of Others (Other representing society and purposely spelt with a capital 'O'). For example, when we are born, we have an unconscious choice to enter into the language of our parents (Cantin, 2002). What they name us, their desires for us, and how they communicate their understanding of the world all contribute to how our knowledge is structured. Our parents' language is interwoven with the broader language (such as laws, norms, cultural meanings) structuring society, which Lacan calls the *Symbolic Order*. The language we come to internalise offers coordinates linking a person to the Symbolic Order that structures interpersonal relations (Hook, 2011).

Yet, in this process, we are presented with a false choice of taking on the Other's language. Lacan compared it to the option given in a robbery: 'Your money or your life', meaning that you may not survive within society if you do not take on the Other's language. Thus, this process of coming into being through the language of Others may be traumatising for some (Cantin, 2002).

This is particularly true if we understand capitalism as its own language ordering the world. Capitalism structures the imaginary register – the realm of images, representations and identifications that form the ego – shaping our sense of self and relationship with others (Lacan, 1988). Capitalism is highly effective in offering us an image of an ideal ego – what we should desire and shaping our identifications in the world (McGowan, 2016). Hyper-individualism, social status, competition, and the accumulation of personal wealth that – we are told – can buy us social standing, and superficial enjoyments are privileged over the mutual flourishing of life on Earth (McGowan, 2016; Žižek, 2009). This logic successfully enchants the upper and middle classes by giving coordinates for what is deemed a successful life.

However, like a snake devouring itself, capitalism rapidly depletes the Earth (Klein, 2015). Capitalism deems more-than-human life – including our own – a commodity that should be exploited for economic production and affirming Western ways of being in the world. It does not regard the planet's ecological boundaries and treats ecosystems like open-access systems to be colonised (Shiva, 2020). One of the most significant sacrifices we make when we come into being through the language of capitalism is that it alienates us from the more-than-human world. Moreover, it insidiously captures our desires and conceals our complicity, successfully keeping the suffering it inflicts on our bodies, minds and the planet out of our purview. Consequently, what is alienated in coming into being through capitalism is challenging to symbolise and is often left unspoken but ready to return.

The global changes associated with capitalism taking form (such as climate change and ecological collapse) are characterised by dismantling structures that bind together different worlds in economic, cultural, social, and ecological dimensions, leading to new forms that remain unknown and undefined (Librett, 2020). For the psychoanalytic subject, this globalising process creates new ways of existing in the world that, in turn, unsettles and disintegrates other ways of knowing and being. These new forms (also referred to as world-formation) present both opportunities and threats to humanity as a whole and, particularly, the psychoanalytic subject.

Based on my clinical work and community psychology practice, many people experiencing climate-related distress have a sense of awakening to and living with the overwhelming significance and urgency of the climate and ecological crises, dealing with the moral dilemmas associated with the personal and societal impacts of living in a capitalist society and grappling with finding sustenance, meaning and a sense of flourishing within capitalism's wreckage and impending suffering. Like others (Librett,

2020), Clarke was presented with an ethical choice about how to respond to these existential threats.

## A Manic Response

Clarke's shared planetary trauma that was so heavily psychically defended against was threatening to return – and it did. Yet, unlike most people I have seen who are more neurotic, Clarke presented with a manic response to the existential, psychic threat. *Don't Look Up* unsettled the perceived solidity of his relationship with the Symbolic Order shattering a sense of security. Simply, he could no longer continue participating in a society through which he had come into being but no longer found tenable.

Practically, Clarke radically began breaking away from his work, spending every minute of the day transitioning his family's lifestyle – changing lightbulbs, installing grey water tanks and cutting out meat – to manage the threat. He would sleep little and, being newly aware of the current mass extinction event, would voraciously read about the catastrophe, which gave him access to new knowledge and only amplified his moral outrage.

In parallel, it seemed that Clarke began unconsciously foreclosing the traditional link between himself and the Symbolic Order. From a Lacanian perspective, it would be understood that the Name-of-the-Father, as a function of authority, law and social order, began de-anchoring, in turn, de-linking the subject of the unconscious from the Symbolic Order. This break appears most clearly in a new framework of understanding that emerged for him. He explained that he was propelled by a new biblical mission, which was confirmed by random signs in the world (such as billboards, symbols and words people said in conversation to him) that he loosely associated with signs in scriptures. These signs also propelled his mission, further elaborating his framework. This was alarming for his family, who described him as 'not being very religious'. Attempts at finding anchorage during psychosis through such religious frameworks, among other frameworks, is common (Fimiani, 2021). However, what was particularly concerning his family and I was that Clarke thought he was the Messiah and that by killing himself, a biblical prophecy would be fulfilled: he would, he thought, save the world.

From a Lacanian perspective, one can theorise that the unsettling of Clarke's links to the Symbolic Order and the disruption of imaginary registers was liberating and unleashed drive energies (Librett, 2020). Librett (2020) explains, consequently, that these drive energies have few possibilities for anchoring in traditional value systems that are losing their solidity and demonstrating an unbearable lack. Through destabilising these registers, the suffering capitalism inflicts on our bodies, minds, and the planet returns as a psychic trauma – also known as *the Real* (Hoornaert, 2022; Librett, 2020). The subject may, in turn, experience a radical break from the symbolic and imaginary registers that usually help to ground a person within the world while also coming under attack from the overwhelming, unsymbolised reality of the Real. The loss of these coordinates to the Other also requires the installation of a new unconscious logic.

Clarke's unbridled energy, loose associations and significations embedded in a hyper-religious framework were hallmarks of this radical break from the symbolic and imaginary registers that took form through a manic-depressive response.

The following clinical aspect of this case formulation draws on Leader's (2013) general description of manic-depressive responses. Broadly, he explains, mania may be followed by a melancholic response, and people may cycle through these experiences. In the structural framework of Lacanian psychoanalysis, manic-depressive responses are understood to be relevant to the Clinic of Psychosis, which is defined by working with foreclosure. It is here where Lacanians have made a significant contribution to psychoanalytic theory by understanding that psychosis is an ethical position where a choice is made to reject links to the Symbolic Order (Fimiani, 2021). Instead of working through the contradictions, clashes and messiness of how the world is becoming, the manic subject psychically splits the world into a world of high contrast, good and bad, love and hate, where there is no overlap, explains Leader. From an adaptive perspective, fantasies, such as his biblical suicidal mission, can play a role in addressing climate change while ameliorating the guilt associated with the crisis and one's despair regarding the climate catastrophe. A quick resolution is available that conceals the complexity of the crisis. This process is similar to what Dodds (2011: 69) describes as 'manic denial'. Moreover, Clarke's sense of vitality and energy, like others experiencing mania, seemed proportional to the loss within his relationship with the symbolic realm. Thus, by foreclosing symbolic links to the logics of capitalism, the manic state likely helped Clarke cope with his threatening existential links to a social world complicit in the death of life on Earth.

Conversely, what also returns in the manic episode is the sense of connectedness with other people and the more-than-human world. Leader (2013: 11) explains: 'A manic episode can give someone a sense of being genuinely alive and connected to the world, of having found one's true identity for the first time. This may be difficult to give up.' Thus, having unanchored from the symbolic realm, connections are made through the language that the manic subject encounters. Links are made rapidly with what Leader describes as an 'irrepressible, brutal persistence'.

Relatedly, I had also been inscribed into Clarke's mission. He would often describe me as part of his crew and someone that could help him – perhaps unconsciously because I am known as a climate-aware psychotherapist. This idealisation of the Other that fits into the altruistic fantasy of saving the world can help stabilise someone working through mania. Leader warns that the functionality of this dynamic should not be readily dismissed.

Speech is also critical. Like Clarke, Leader explains that the exhilaration of the sense of connectedness must be communicated. People experiencing mania differ from other subjects of psychosis who instead may withdraw into paranoia that requires limited to no social link with others. While someone with paranoid psychosis may thus enjoy their knowledge alone, the person with manic depression must urgently share this knowledge with the world (Leader, 2013). In mania, speech tends to revert to the same ideas, words or significations and the interlocutor of the social link to whom speech is addressed is prevented from any real reply. The manic subject can then be thought of as the knowing subject looking for an address to deliver a particular message to another (Fimiani, 2021). For Clarke, 'the mission', 'saving the world', 'impending crises', 'people needing to wake up' all became repetitive slogans.

In response to these dynamics, the Lacanian-oriented therapist takes an ethical position by playing the role of a witness, detailing and reflecting on what is described, and, over time, linking the subject again in some way to a symbolic realm (Fimiani, 2021; Fink, 2011). Moreover, over time, the witnessing analyst works sensitively as a listener towards the edges and doubts in the certainty and *the* truth of the logical framework (Fimiani, 2021). This dialogue, in turn, also allows for what underlies the manic response to be spoken, slowly but surely creating new ways for such experiences to be understood and, sometimes, worked through. Consequently, this process helped unsettle the certainty Clarke had in his self-sacrificing mission.

This method starkly contrasts conventional biomedical approaches to a diagnosis of bipolar or psychosis that deem such experiences medically based, delusional and unsafe to discuss owing to psychiatrists' fear of worsening symptoms (Fimiani, 2021). Moreover, therapists aligned with such biomedical ideologies are directed to challenge beliefs as meaningless productions of what is erroneously considered a chemically unbalanced mind. The Lacanian approach vehemently opposes such dehumanising and fear-based perspectives that close down speech and subjectivity.

Even though Clarke did opt to take mood stabilisers, people he encountered in the mental healthcare system shut down his fears about the climate crisis and attributed his experience solely to a psychiatric disorder. Closing down his speech at this point only hastened the movement towards paranoid thinking about such care. Leader (2013) explains that when speech difficulties are experienced, and dialogue falls flat, a person may experience paranoia as anxiety escalates. Simply, the mental health system becomes part of a threatening authority that is also unceasingly tolerated or rejected outright. In contrast, working through the magnitude of the confrontation with the climate crises can serve as an act of witnessing and re-linking to the social world.

A manic response comes about as a call to meaning where the urgency, altruism and sacrificial logic underpin disorganised associations and grand theories about the world. These grandiose themes tend to enlarge the image of oneself and place the Other in a position of ultimate good, protecting against the horror of the climate crises and the Symbolic Order's loss of solidity. As Leader (2013: 43) explains: 'A giving and generous world must be preserved at all costs.' However, this grandiose, inflated sense of altruism in these states also poses a risk. In the case study, Clarke was willing to pay the ultimate price as a sacrifice through his death, which he perceived would ultimately save the world. For example, splitting may be stabilising in that the 'bad' is attributed to the Other that is seen to be outside of oneself and, consequently, a clear us against them scenario is manifested, Leader explains. However, splitting can make it difficult for the person to symbolise the nuances of the situation.

A critical position then was for me to help reflect some of the feelings that he was experiencing, as he relayed them to me, and create space for working through. This was no easy feat as Clarke would sometimes become reasonably overwhelmed by the subject of the catastrophe, albeit coming to these topics on his own. Conversely, when we veered off the subject and focused on his wellbeing, he became visibly agitated. He wanted to speak.

In the process, I had to sit through my own fears. For example, I was concerned about how Clarke's overwhelming sense of dread might amplify the unrelenting urgency for his suicidal, altruistic fantasy. These fears must be checked, and are often unfounded, consequently closing down speech (Fimiani, 2021). As therapists, we must defend against uninterrogated worries that prevent us from meaningfully engaging with topics that understandably cause distress. Fears about triggering mania can do damage, leaving the person feeling alienated and alone to work through troubling experiences.

Nevertheless, during mania, the therapist's insights, associations and inputs may be limited in their effect. Leader (2013) posits that this is owing to words being unanchored in the fullness of their meaning. Instead, therapy can play an essential role in helping to symbolise some of the rejected complexity. To do so, I created a therapeutic space to speak, taking breaks and doing breathing exercises that Clarke found reduced some of his anxiety at its peaks. This process showed that he could work through these distressing thoughts, grounding him in the moment. As the mania resolved, I also worked with him to reconnect to other areas of his life, including having more involvement with his family, body and work.

However, the resolution of the manic state created another set of challenges. While Clarke's manic response defended against an overwhelming loss and brought a grand sense of connection, the despair of his complicity in the climate crisis, like that of many others, returned during the melancholic stage.

## Melancholia

Although words are unanchored in their symbolic links in mania, space is opened in melancholia. Here, too, the therapist plays the role of a witness

(Hook, 2022). The cost of this opening up is that words are scarce and weighed down by the gravity of what is signified – their concrete meanings (Leader, 2013). In contrast, Clarke's depressive speech in the aftermath of the manic response was characteristically impoverished, significations were self-critical, and suicidality became more evident.

Leader (2013) explains that the significations of a depressive state include being worthless, a sense of being spiritually void or being guilty of some terrible ineradicable sin. In this state, the weight of society's complicity in the climate crisis and daily non-action became a likely tool for self-punishment. Clarke positioned suicidality as an inevitable choice. Thus, while a mania may foreclose the sense of guilt completely, it is likely to return brutally in the experience of melancholia. We continued reconnecting places of meaning, including his relationship with family, as it was his family's suffering in relation to climate change that also caused him profound distress, yet also gave him the will to live. Also, instead of discounting his despair and meaninglessness towards his work as a banker, we talked through Clarke's desire to play a more active role in the climate justice struggle – particularly the scaling up of solar as part of the just transition, a movement that aims to safeguard workers and communities through the creation of an equitable zero-carbon economy. Practically, for Clarke, this involved exploring concrete opportunities. The most remarkable shift in his clinical process was after he found a job in the renewable sector, where he found new coordinates to re-member himself within the Symbolic Order.

## Conclusion

This case study has portrayed someone's manic-depressive response to the overwhelming reality of the climate crises through a Lacanian psychoanalytic lens. To my knowledge, this is the first documented case study of manic-depressive responses to the climate crisis. By integrating Lacanian psychoanalysts, such as Fimiani and Leader, towards psychosis and manic-depressive responses, and Apollon and colleagues concerning psychosis and globalisation, this chapter has presented insights for working with manic-depressive responses in the age of climate breakdown. When the content of mania or psychosis is related to climate change, fear of triggering another episode seems to close necessary dialogue about the experience of climate change, including an overwhelming fear and moral urgency to act. More research involving those that have experienced manic-depressive responses is required.

Several other non-pathologising psychotherapeutic models such as the Power Threat Meaning Framework (Ball, Morgan and Haarmans, 2023; Morgan et al., 2022) and peer-support models such as the Hearing Voices Network (2023) may also be useful in identifying how to support those who desire it. In this case study, Clarke held the significance and urgency of the moment in defiance of how desensitised and non-reactive the world around him was (and still is). Psychosis, as a position, has long been considered a subversive force for social change (Deleuze and Guattari, 2009) and can cut through the denial of the moment. Ultimately, therapy provided Clarke the space to speak through the unsettling experience, which enabled his re-linking to the Symbolic Order.

## Note

1 I do not suggest that there should be a diagnosis of 'climate mania' but rather a recognition of manic-depressive responses to the climate crisis.

## References

Apollon, W., Bergeron, D., and Cantin, L. (2002) *After Lacan: Clinical Practice and the Subject of the Unconscious*. New York: State University of New York Press.

Ball, M., Morgan, G. and Haarmans, M. (2023) The Power Threat Meaning Framework and 'Psychosis'. In: J. A. Díaz-Garrido, R. Zúñiga, H. Laffite et al. (eds), *Psychological Interventions for Psychosis*. New York: Springer International Publishing, pp. 141–169. 10.1007/978-3-031-27003-1_8.

Cantin, L. (2002) The Trauma of Language. In W. Apollon, D. Bergeron and L. Cantin (eds.) *After Lacan: Clinical Practice and the Subject of the Unconscious* (First edition), pp. 35–48. New York: State University of New York Press.

Deleuze, G. and Guattari, F. (2009) *Anti-Oedipus: Capitalism and Schizophrenia*. London: Penguin Classics.

Dodds, J. (2011) *Psychoanalysis and Ecology at the Edge of Chaos: Complexity Theory, Deleuze, Guattari and Psychoanalysis for a Climate in Crisis*. Oxon: Routledge.

Fimiani, B. (2021) *Psychosis and Extreme States*. New York: Springer International Publishing. 10.1007/978-3-030-75440-2.

Fink, B. (2011) *Fundamentals of Psychoanalytic Technique: A Lacanian Approach for Practitioners*. New York: Norton Professional Books.

Hearing Voices Network. (2023) *Hearing Voices Network*. Available at: www.hearing-voices.org/ [Accessed 26 July 2023].

Hook, D. (2011) Empty and Full Speech. In: D. Hook, B. Franks and M. W. Bauer (eds) *The Social Psychology of Communication*. London: Palgrave Macmillan. 10.1057/9780230297616_10.

Hook, D. (2022). The complex of melancholia. In D. Hook and S. Vanheule (eds) *Lacan on Depression and Melancholia*. Oxon: Routledge, pp. 104–119. 10.4324/9781003216391-8.

Hoornaert, G. (2022, June 29). *Carbon Blitz*. The Lacanian Review. Available at: www.thelacanianreviews.com/carbon-blitz/ [Accessed on 13 May 2023].

Intergovernmental Panel on Climate Change (IPCC) (2023). *(2022) Climate Change 2022: Impacts, Adaptation, and Vulnerability. Contribution of Working Group II to the Sixth Assessment Report of the Intergovernmental Panel on Climate Change*. Cambridge: Cambridge University Press. doi:10.1017/9781009325844.

Klein, N. (2015) *This Changes Everything: Capitalism vs. the Climate*. New York: Simon and Schuster.

Lacan, J. (1988) In J. A. Miller (ed.) *The Seminar of Jacques Lacan, Book 1: Freud's Papers on Technique (English and French edition)*. New York: Norton Professional Books.

Leader, D. (2013) *Strictly Bipolar*. London: Penguin.

Librett, J. S. (2020) The Subject in the Age of World-formation (Mondialisation): Advances in Lacanian theory from the Québec Group. In: A. Govrin and J. Mills (eds) *Innovations in Psychoanalysis: Originality, Development, Progress*. London: Routledge, pp. 75–99.

McGowan, T. (2016) *Capitalism and Desire: The Psychic Cost of Free Markets*. New York: Columbia University Press.

Meyer, C. (2023) The Logic of Fantasy in the Direction of the Treatment and the Analytic Act [seminar]. *Lacan School of Psychoanalysis*.

Morgan, G., Barnwell, G., Johnstone, L. et al. (2022) The Power Threat Meaning Framework and the climate and ecological crises. *Psychology in Society (PINS)*, 63: 83–109.

Shiva, V. (2020) *Reclaiming the Commons: In Defense of Biodiversity, Traditional Knowledge and the Rights of Mother Earth*. Santa Fe: Synergetic Press.

Žižek, S. (2009) *The Sublime Object of Ideology*. New York: Verso.

Chapter 12

# Climate Sorrow: Discerning Various Forms of Climate Grief and Responding to Them as a Therapist

*Panu Pihkala*

## Introduction

Grief is a complex issue in contemporary societies. Fundamentally, grief and sadness are related to changes and especially aspects of them that are felt as losses. As such, they are universal human experiences that are happening everywhere. Various forms of grief, sadness and sorrow are often present in therapeutic encounters of any kind.

This universality is contrasted by the complexities and inadequacies in industrialised cultures in relation to grief. Frequently, grief is feared and even pathologised (Horwitz and Wakefield, 2007). Westernised social norms often support displays of contentment, not displays of sadness (Ehrenreich, 2009); ancient human practices around grief and mourning tend to be marginalised or even forgotten, for example, wearing certain clothes for a period of mourning or visiting those members of a community who are grieving (for further discussion of this, see Menning, 2017). Furthermore, some forms of grief are psychosocially disavowed, a phenomenon which has been described as disenfranchised grief (Doka, 1989).

Grief and sadness arising out of ecological concerns has been a difficult topic to engage with partly because of the general difficulties around grief, described above, and partly because of the particular difficulties around environmental issues. Ecological grief is psychologically hard to bear at an individual level and can manifest in different ways such as denial or depression (Kidner, 2007; Weintrobe, 2021). Different values and worldviews also contribute; for some people, clearcutting an old forest is a traumatic loss, while for others it means economic gain and positive development. Ecological grief is often 'disenfranchised grief' (Cunsolo Willox and Landman, 2017; Cunsolo Willox and Ellis, 2018), and those who feel it have sometimes felt misunderstood even by therapists (Stoknes, 2015). Therapists have opportunities to support people who are experiencing ecological grief. However, this requires self-reflection and preparation.

DOI: 10.4324/9781003436096-24

## Ecological Grief and Climate Grief

Climate grief is defined here as part of the broader phenomenon of ecological grief or 'environmental grief' (Kevorkian, 2020), which is significantly related to the climate crisis. Differentiating between climate grief and other forms of ecological grief can be difficult because climate change affects so many other kinds of ecological loss and grief, but it is important to make distinctions between these concepts. The various forms of ecological grief need recognition; exploring and validating these various forms is a crucial task for therapeutic encounters (Doherty, 2018; Gillespie, 2020; Weber, 2020; Pihkala, 2022a).

Some people recognise that they are experiencing climate grief. Others may feel it without being fully aware of it, yet it lives in their bodies. It is difficult not to be affected in some way by climate grief in the contemporary world, because there are both ecological and social losses generated by the climate crisis. Even if people do not have a strong connection to the environmental or profound environmental values, they are affected by the impacts on societies and communities, if only unconsciously. In therapeutic encounters, climate grief can manifest in many ways – some people seek direct support, while others may wrestle with it and struggle to put their experience into words.

Climate grief can be an acute, bereavement-like reaction after awakening to the severity of the climate crisis (Randall, 2009; Pihkala, 2022b). It can also manifest as a mood, like sadness, which can be persistent. Among young people, it has been observed that they may function normally but still be haunted by the background presence of the climate crisis and related grief (Diffey et al, 2022).

Sometimes climate grief turns into melancholy – a persistent condition where self-esteem suffers and expectations are low. For more on environmental melancholia, see Lertzman (2015). Some scholars have explored the term 'climate depression' (Budziszewska and Kałwak, 2022), which in my mind could be defined as depression-related phenomena that are significantly connected with the climate crisis. It is unnecessary to diagnose or create a DSM/ICD categorisation; it is enough to observe the role of the climate crisis in these states where such descriptions might be created, such as Adjustment Disorder or complicated grief (Doherty, Lykins and Piotrowski, 2021).

Together with other scholars and psychologists, I would advocate against pathologising climate grief, as difficult as it can be to hold such feelings in a therapeutic setting. Climate grief is a rational response to the reality of climate breakdown. This crisis will last for generations, and thus feelings of sadness may become constant companions.

Fundamentally, sadness helps people to engage with losses and to cherish what is valuable to them (Greenspan, 2004; Solomon, 2004; McLaren, 2010; Lomas, 2016). It is an emotion that assists people in reconstructing their lives amidst changes and losses. Thus, the therapist encounters a challenging but highly important balancing task: to pay attention to possible clinically significant manifestations of climate grief, but fundamentally to have a sadness-affirming attitude.

## Aspects of Climate Grief

An important question to ask is: What is the person grieving when they feel climate grief?

Although various issues in a person's life can affect climate grief, therapists should not begin with the presupposition that climate grief is mostly generated by other psychological and social factors (Stoknes, 2015) It does, however, need to be considered how aspects of a person's experience and life history may re-emerge and interrelate with climate grief (Haseley, 2019).

Table 12.1 shows the many overlapping dimensions of climate grief; for example, when the dimension is ecological, there is grief for changes in the ecosystems. See also Crandon and colleagues (2022) who use a socio-ecological framework to explore the different factors influencing the experience of climate anxiety adapted here for grief. The therapist may find it helpful to refer to the dimensions listed in this table to map and inform understanding of the client's experience of climate grief. There can be complex dynamics between the global and the local which may interact together in profound ways (Bodnar et al., 2022).

As climate change is a planetary phenomenon, climate grief inevitably has a global character and usually manifests as a universal type of grief in the person's life. It is related to the person's worldview, to their entire lifeworld. It is a kind of world grief, indeed *weltschmerz* in a literal form. In this sense, it is closely connected with what is commonly called climate anxiety or climate distress (Pihkala, 2020a). The character of climate change as a Hyperobject (Morton, 2013) complicates matters: climate change is everywhere and nowhere at the same time.

Alongside the global, climate grief can also be a response at a local level to the various losses an individual and their community experience. For

*Table 12.1* Dimensions of Climate Grief and what may be grieved in each dimension

| *Dimensions* | *We Mourn Changes In* |
|---|---|
| Ecological | ecosystems, landscapes, plants, wildlife etc and familiar natural environment |
| Self | one's image of self in relation to others, status and place in society, loss of sense of self |
| Close relations | close relationships near and far, family, friends and animal companions |
| Social world | community and wider society |
| Cultural level | cultural roots – traditions and belonging |
| Global level | the global environment and culture |

place-related grief, the concept of 'solastalgia' helps to locate the loss (Albrecht et al., 2007).

## Guilty Grieving

Scholars and psychologists have noted that climate grief is 'guilty grieving' (Menning, 2017; Jensen, 2019). While people's experiences of guilt and shame differ, it is vital to pay attention to this dimension in any work with climate grief, particularly as there may be a complex interplay with the therapists' own feelings of climate guilt.

In any process of grief, there is a dimension of engaging with guilt. This is especially difficult if there is clear culpability involved in the loss (Worden, 2018). People experiencing grief may find themselves asking questions related to guilt and responsibility. The question 'Should I have done something differently?' can turn into a ruminating and haunting belief of what 'I *should* have done differently'. Often there are hopes and regrets involved: 'If I had only known, I would have acted differently …' (Li et al., 2014).

In climate grief, the dynamics of guilt and responsibility are highly complex because the climate crisis is interconnected with the whole contemporary way of life, particularly in the Global North. Carbon emissions are embedded in all aspects of our lives. It is difficult to navigate between taking personal responsibility and succumbing to the weight of guilt or shame. Social and political factors can amplify or reduce responses. For example, the public relations campaigns of oil companies have intentionally tried to make individuals feel more guilt in order to defuse the emotional energy of righteous anger demanding changes (Jensen, 2019).

If and when people engage with climate reality, grief and guilt dynamics can be especially powerful especially for those who live in high-income countries. They can experience a combination of colonial guilt and survivor guilt because others are suffering more acutely from climate breakdown (Davenport, 2017). As Jensen (2019) observes, climate grief intensifies climate guilt, and climate guilt complicates climate grief. One important therapeutic task is to recognise possible manic responses to climate guilt and grief, for example immersing oneself so intensively into climate action that resources run out. For climate burnout see Hoggett and Randall (2018).

The combined dynamics of climate grief and guilt are one reason why therapists can find climate so difficult to engage with. Therapists often have climate guilt themselves, and if they are wealthy, there may be complex dynamics of guilt and privilege. The challenge of living with the dialectics of climate guilt is thus a joint challenge for both the therapist and the client, and many kinds of transferences and defences may develop between them in therapeutic encounters (Lewis, Haase and Trope, 2020; Silva and Coburn, 2022).

Scholars of grief recommend reality-testing guilt (Worden, 2018), and this would also be valuable in relation to climate guilt. In what ways is the guilt possibly valid? In what ways may it be exaggerated? One important therapeutic tool here is dialectical thinking, through which one can accept both individual responsibility and the greater responsibility of big actors and structures (Lewis, Haase and Trope, 2020).

## Grief and Bereavement Theory as Applied to Climate Grief

### *Mediators of Mourning*

Grief theory may help to discern various factors that have a role in experiences of ecological and climate grief. Worden's framework of mediators of mourning (2018, Ch 3) can be helpful here, although dynamics differ if the loss does not concern the death of someone close. Mediators include personality and social variables, coping style, historical antecedents of grief, nature of attachments with objects of grief and concurrent losses and stresses (Comtesse et al., 2021). Worden's questions 'Who died?' can be applied to climate grief by formulating them as 'What is felt to be lost' and 'How did these losses happen?'.

### *Temporalities*

Climate grief frequently focuses on the loss of an anticipated future, in a similar way to climate anxiety and worry. However, all temporalities are at play with climate grief. People may mourn losses that have already happened due to climate change. What is happening at the present moment affects people strongly but also has impacts on predictions and feelings about the future (see Barnett, 2021).

Cunsolo Willox and Ellis (2018) discuss the phenomenon of anticipatory grief/mourning in relation to climate grief. For practical therapeutic work, two aspects are particularly significant: bearing witness to catastrophising and being open to the possibility that the losses will increase with time.

### *Tangible and Intangible Losses*

Harris (2020) provides a helpful discussion of tangible and intangible losses. Tangible losses refer to losses that are perceived with the senses. Intangible losses are invisible or at least difficult to notice. Both are important to explore in therapeutic encounters (Tschakert et al., 2019).

Some intangible climate-related losses are more difficult to notice than others. These include the loss of dreams or future hopes. Some losses have both tangible and intangible aspects, such as the loss generated by a decision not to have children because of the climate crisis. The tangible aspect is the child that is not there, and intangible aspects can include the loss of the role of parent or grandparent. For people with a close relationship with the land, such as Indigenous peoples, climate change produces a huge complex of tangible and intangible losses; their whole identity and worldview are under pressure (Cunsolo Willox and Ellis, 2018).

### *Ambiguous Loss, Frozen Grief, Nonfinite Loss and Chronic Sorrow*

Ambiguous loss refers to losses where there is either uncertainty about the exact state of the loss or a situation where something is partly lost and partly not; a classic example is the mourning of soldiers missing-in-action. Climate change produces many ambiguous losses; seasonally snowy winters may be partly gone and partly not.

Nonfinite loss refers to losses that have no foreseeable end. In general grief research, a classic example of such a condition would be a permanently disabling injury. The loss may generate other losses and there may be much uncertainty and feelings of helplessness. This may result in feelings of being outside the mainstream social world (Harris, 2020), a form of psychological isolation. This concept suits climate grief in numerous ways, since the negative impacts of climate change will continue indefinitely and people who feel climate grief have often felt socially isolated or marginalised.

Frozen grief describes the situation when grief gets stuck (Boss, 2020). Grief and sadness can arise constantly but irregularly; the concept of chronic sorrow can be useful in understanding this (Ross, 2020). There is often disenfranchisement of chronic sorrow, ambiguous loss and nonfinite loss, and since climate grief can have all these dynamics, there is a strong need to validate and recognise such losses and forms of grief.

### *Meaning Reconstruction*

One final aspect of grief theory to consider is the meaning reconstruction framework of Neimeyer and colleagues (Neimeyer, 2019). This integrative framework uses narrative methods to explore the ways in which systems of meaning undergo changes in the processes of grief. My view is that the process of engaging with the changing socio-ecological state of the world entails deep meaning reconstruction and a task of "re-learning the world", as grief philosopher Thomas Attig describes (Pihkala, 2020b).

## Climate Grief in the Therapy Room

Bringing together the ideas in this chapter, recommendations and ideas for therapeutic encounters are listed below.

- Nurture an attitude that understands the complexities of climate grief but does not pathologise the phenomenon, even while keeping an eye for clinically significant manifestations of grief, anxiety and depression

- A sadness-positive attitude can help to understand how grief and sadness serve life
- Utilise many insights of general grief literature but be also open to special features of climate grief
- Therapists' own climate emotions can affect therapeutic encounters in profound ways as we all are inside climate change – therapists need to mindful of this in engaging with climate grief
- Make use of resources provided by climate-aware therapy and eco-therapy literature (e.g. Doherty, 2018; Rust, 2020; Aspey, Jackson and Parker, 2023)
- Analyse various felt losses that people may experience; many losses related to the climate crisis are intangible and may be difficult to identify
- Observe possible impacts of disenfranchised grief and ambiguous loss
- Take note that climate grief may include complex processes in relation to self-world relationships, touching on emotional attachments and parts of the self
- Explore other climate emotions that may be intertwined with climate grief and sadness (Pihkala, 2022b) perhaps using mind-mapping
- Apply process models and/or other graphic models (e.g., Pihkala, 2022a; Hickman, 2023) to explore people's journeys with climate grief and other emotions
- Be open to the long-running process of meaning reconstruction that climate grief entails
- Validate also the possibilities of experiencing joy and contentment amidst processes of climate grief
- Understand that climate grief which may include responses to nonfinite losses. You might consider the clinical recommendations of (Schultz and Harris, 2011) on nonfinite loss:

    - Name and validate the loss(es)
    - Normalise the ongoing nature of the loss
    - Find support and resources
    - Recognise the loss(es) and identify what is not lost
    - Allow for the possibility of meaning making and growth
    - Initiate rituals where none exist

Finally, it is important to acknowledge both the tragedy and the care that are present in climate grief. It is difficult to live through ecological breakdown, but everyday manifestations of empathy, such as those happening in therapeutic encounters, testify to a 'culture of care' (Weintrobe, 2021) which provides meaning.

## References

Albrecht, G., Sartore, G-M., Connor, L. et al. (2007) 'Solastalgia: The Distress Caused by Environmental Change', *Australasian Psychiatry: Bulletin of the Royal*

*Australian and New Zealand College of Psychiatrists*, 15(S1): S95–S98. 10.1080/10398560701701288.

Aspey, L., Jackson, C. and Parker, D. (eds) (2023) *Holding the Hope: Reviving Psychological and Spiritual Agency In the Face of Climate Change*. Monmouth: PCCS Books.

Barnett, J. T. (2021) Vigilant Mourning and the Future of Earthly Coexistence. In A. M. Dare and C. Vail Fletcher (eds) *Communicating in the Anthropocene: Intimate Relations*. Lanham: Lexington Books, pp. 13–33.

Bodnar, S., Aliovin, P., O'Neill, P. et al. (2022) The Environment as an Object Relationship: A Two-part Study, *Ecopsychology*. 10.1089/eco.2022.0070.

Boss, P. (2020) Understanding and Treating the Unresolved Grief of Ambiguous Loss: A Research-based Theory to Guide Therapists and Counselors. In D. L. Harris (ed.) *Non-Death Loss and Grief: Context and Clinical Implications*. New York: Routledge, pp. 73–79.

Budziszewska, M. and Kałwak, W. (2022) Climate Depression: Critical Analysis of the Concept, *Psychiatria Polska*, 56(1): 171–182. 10.12740/PP/127900.

Comtesse, H., Ertl, V., Hengst, S. M. C. et al. (2021) Ecological Grief as a Response to Environmental Change: A Mental Health Risk or Functional Response?, *International Journal of Environmental Research and Public Health*, 18(2): 734. 10.3390/ijerph18020734.

Crandon, T. J., Scott, J. G., Charlson, F. J. et al. (2022) A Social-ecological Perspective on Climate Anxiety in Children and Adolescents, *Nature Climate Change*, 12: 123–131. 10.1038/s41558-021-01251-y.

Cunsolo Willox, A. and Ellis, N. R. (2018) Ecological Grief as a Mental Health Response to Climate Change-related Loss, *Nature Climate Change*, 8(4): 275–281.

Cunsolo Willox, A. and Landman. K. (eds) (2017) *Mourning Nature: Hope at the Heart of Ecological Loss & Grief*. Montreal & Kingston: McGill-Queen's University Press.

Davenport, L. (2017) *Emotional Resiliency in the Era of Climate Change: A Clinician's Guide*. London: Jessica Kingsley Publishers.

Diffey, J., Wright, S., Uchendo, J. O. et al. (2022). 'Not About Us Without Us': the Feelings and Hopes of Climate-concerned Young People Around the World. *International Review of Psychiatry* 34(5): 499–509. 10.1080/09540261.2022.2126297.

Doherty, T., Lykins, A. D. and Piotrowski, N. A. (2021) Clinical Psychology Responses to the Climate Crisis. *Reference Module in Neuroscience and Biobehavioral Psychology*. 10.1016/B978-0-12-818697-8.00236-3.

Doherty, T. J. (2018) Individual Impacts and Resilience. In: S. D. Clayton and C. M. Manning (eds) *Psychology and Climate Change: Human Perceptions, Impacts, and Responses*. Amsterdam: Academic Press, pp. 245–266.

Doka, K. J. (1989) *Disenfranchised Grief*. Lexington, MA: Lexington Books.

Ehrenreich, B. (2009) *Bright-sided: How the Relentless Promotion of Positive Thinking Has Undermined America*. New York: Metropolitan Books.

Gillespie, S. (2020) *Climate Crisis and Consciousness: Re-imagining Our World and Ourselves*. Oxon: Routledge.

Greenspan, M. (2004) *Healing Through the Dark Emotions: The Wisdom of Grief, Fear, and Despair*. Boulder: Shambhala.

Harris, D. L. (ed.) (2020) *Non-Death Loss and Grief: Context and Clinical Implications*. Oxon: Routledge.

Haseley, D. (2019) Climate Change: Clinical Considerations, *International Journal of Applied Psychoanalytic Studies*, 16: 109–115. 10.1002/aps.1617.

Hickman, C. (2023) Feeling Okay with Not Feeling Okay: Helping Children and Young People Make Meaning from Their Experience of Climate Emergency. In: L. Aspey, C. Jackson and D. Parker (eds) *Holding the Hope: Reviving Psychological and Spiritual Agency in the Face of Climate Change*. Monmouth: PCCS Books, pp. 183–198.

Hoggett, P. and Randall, R. (2018) Engaging with Climate Change: Comparing the Cultures of Science and Activism. *Environmental Values*, 27(3): 223–243. 10.3197/096327118X15217309300813.

Horwitz, A. V. and Wakefield, J. C. (2007) *The Loss of Sadness: How Psychiatry Transformed Normal Sorrow into Depressive Disorder*. Oxford: Oxford University Press.

Jensen, T. (2019) *Ecologies of Guilt in Environmental Rhetorics*. Cham: Palgrave Macmillan.

Kevorkian, K. A. (2020) Environmental Grief. In D. L. Harris (ed.) *Non-Death Loss and Grief: Context and Clinical Implications*. Oxon: Routledge, pp. 216–226.

Kidner, D. W. (2007) Depression and the Natural World: Towards a Critical Ecology of Psychological Distress, *Critical Psychology*, (19): 123–146.

Lertzman, R. A. (2015) *Environmental Melancholia: Psychoanalytic Dimensions of Engagement*. Oxon: Routledge.

Lewis, J., Haase, E. and Trope, A. (2020) Climate Dialectics in Psychotherapy: Holding Open the Space between Abyss and Advance, *Psychodynamic Psychiatry*, 48(3): 271–294.

Li, J., Stroebe, M., Chan, C. L. W. et al. (2014) Guilt in Bereavement: A Review and Conceptual Framework, *Death Studies*, 38(1–5): 165–171. 10.1080/07481187.2012.738770.

Lomas, T. (2016) *The Positive Power of Negative Emotions: How Harnessing Your Darker Feelings Can Help You See a Brighter Dawn*. London: Piatkus.

McLaren, K. (2010) *The Language of Emotions: What Your Feelings Are Trying to Tell You*. Boulder: Sounds True.

Menning, N. (2017) Environmental Mourning and the Religious Imagination. In: A. Cunsolo Willox and K. Landman (eds) *Mourning Nature: Hope at the Heart of Ecological Loss & Grief*. Montreal & Kingston: McGill-Queen's University Press, pp. 39–63.

Morton, T. (2013) *Hyperobjects: Philosophy and Ecology after the End of the World*. Minnesota: University of Minnesota Press.

Neimeyer, R. A. (2019) Meaning Reconstruction in Bereavement: Development of a Research Program, *Death Studies*, 43(2): 79–91. 10.1080/07481187.2018.1456620.

Pihkala, P. (2020a) Anxiety and the Ecological Crisis: An Analysis of Eco-anxiety and Climate Anxiety, *Sustainability*, 12(19): 7836. 10.3390/su12197836.

Pihkala, P. (2020b) Climate Grief: How We Mourn a Changing Planet, *BBC Website, Climate Emotions series*. Available at: www.bbc.com/future/article/20200402-climate-grief-mourning-loss-due-to-climate-change? [Accessed 27July 2023].

Pihkala, P. (2022a) The Process of Eco-anxiety and Ecological Grief: A Narrative Review and a New Proposal, *Sustainability*, 14(24): 16628. 10.3390/su142416628.

Pihkala, P. (2022b) Toward a Taxonomy of Climate Emotions, *Frontiers in Climate*, 3. 10.3389/fclim.2021.738154.

Randall, R. (2009) Loss and Climate Change: The Cost of Parallel Narratives, *Ecopsychology*, 1(3): 118–129.

Ross, S. (2020) Chronic Sorrow. In: D. L. Harris (ed.) *Non-Death Loss and Grief: Context and Clinical Implications*. Oxon: Routledge, pp. 192–204.

Rust, M.-J. (2020) *Towards an Ecopsychotherapy*. London: Confer Books.

Schultz, C. L. and Harris, D. L. (2011) Giving Voice to Nonfinite Loss and Grief in Bereavement. In: R. A. Neimeyer, D. L. Harris, H. R. Winokur et al. (eds) *Grief and Bereavement in Contemporary Society: Bridging Research and Practice*. Oxon: Routledge, pp. 235–245.

Silva, J. F. B. and Coburn, J. (2022) Therapists' Experience of Climate Change: A Dialectic Between Personal and Professional, *Counselling and Psychotherapy Research [Preprint]*. 10.1002/capr.12515.

Solomon, R. C. (2004) On Grief and Gratitude. In: R. C. Solomon (ed.) *In Defense of Sentimentality*. Oxford: Oxford University Press, pp. 75–107. 10.1093/019514550X.003.0004.

Stoknes, P. E. (2015) *What We Think About When We Try Not to Think About Global Warming: Toward a New Psychology of Climate Action*. White River Junction: Chelsea Green Publishing.

Tschakert, P., Ellis, N. R., Anderson, C. et al. (2019) One Thousand Ways to Experience Loss: A Systematic Analysis of Climate-related Intangible Harm from Around the World. *Global Environmental Change*, 55: 58–72. 10.1016/j.gloenvcha.2018.11.006.

Weber, J. A. (2020) *Climate Cure: Heal Yourself to Heal the Planet*. Woodbury: Llewellyn Publications.

Weintrobe, S. (2021) *Psychological Roots of the Climate Crisis: Neoliberal Exceptionalism and the Culture of Uncare*. New York: Bloomsbury.

Worden, J. W. (2018) *Grief Counselling and Grief Therapy: A Handbook for the Mental Health Practitioner*. 5th edition. New York: Springer.

Chapter 13

# Coming to Our Senses: Turning Towards the Body

*Tree Staunton*

## Introduction

The work of therapy is to turn crisis into meaning. There is no question that we are living within various overlapping and unfolding crises – social, cultural and environmental.

This chapter explores how far the profession has come in terms of training and regulation that is fit for purpose in a critically endangered world and considers how much further we need to go in our thinking, training and practice. In terms of practice, I will argue that it is the loss of our embodied relationship with ourselves and our world that contributes to our failure to grasp the reality of the interlocking crises we face. I call for a phenomenological approach to all therapeutic explorations, seeking a return to embodied knowing. I consider whether all psychotherapy should be body-oriented if we are to come to our senses in a time of environmental breakdown.

## Developments Within the Profession

In 2015 the Humanistic and Integrative College of UKCP[1] (over 50% of its 11,000 members) created *Environmental Sustainability and Climate Change Guidelines* which were adopted by all training organisations within the College. They were intended as consciousness raising, to form links in the minds of trainees and therapists, stating:

> *To be a psychologically aware human being in our society today means to embrace the ecological interconnectedness of all things and to know with every breath we take that we are dependent on this living system of our environment.*

Our guidelines outlined principles that embody a systemic approach to engaging with the earth and environmental issues offering proposals for ongoing sharing of expertise and knowledge, stating that essentially as

DOI: 10.4324/9781003436096-25

Humanistic organisations, we see that the ecological crisis stems from one of *consciousness* and *relationship*:

> *Psychotherapists have an important role to play in this exploration, enabling a self-reflective dialogue regarding our relationship to the world around us, and our responsibility to it. From this dialogue and reflection, sustainable actions can potentially emerge.*

Dialogue continued between the training organisations within the Humanistic & Integrative College of UKCP and in 2019 myself and colleagues in the Training Standards Committee proposed an addition to the modality-specific training standards to include as a guiding principle *the recognition of our interdependence on our environment and the natural world.* Finally agreed at the end of 2022, these form additions to the compulsory learning outcomes of *sustainability and environmental awareness* in the appendix of the Training Standards *Context of Practice.*[2] Going forward this makes it a training *requirement* for all UKCP-HIPC training organisations to include input such that future qualifying therapists will be able to demonstrate:

- awareness that our current context of practice includes environmental breakdown
- understanding of the implications and likely impacts this will have on the lives and wellbeing of us and future generations
- awareness of the Western European bias in the theory and culture of psychotherapy which impacts on issues of social justice, especially within our current crisis of environmental sustainability
- critical understanding of the unconscious processes which affect our ability to work with and process the impacts of environmental breakdown on ourselves and our clients
- understanding of our environmental identity and how a socially constructed sense of self can create prejudice and discrimination against other life forms
- a recognition of the systemic issues involved in the current environmental crisis, such that the different parts of the system create polarisations and also live within us as individuals

These extra training requirements were fully supported within the Humanistic and Integrative College, and students within my own training organisation expressed excitement at the proposed compulsory addition of a three-weekend module on *The Ecological Self and the Environmental Crisis.*

However, the proposals raised questions amongst some colleagues within different modality approaches. Would this mean a requirement to add climate science to our already packed curriculum? What was the relevance to psychotherapy? Did this represent the addition of a 'political' agenda to psychotherapy training? Whilst the connections between psychological health and environmental breakdown were obvious to me, they were not to my colleagues. I decided to explore attitudes among students and practitioners.

## Focus Group Research[3]

I set up a focus group exploration within my own training organisation with voluntary participants – trainee and graduate therapists – to discuss their feelings about climate change and how it preoccupies them, and to consider what training they feel they need in order to face these issues themselves and with their clients.

Feeling responses were easily accessible and most were in touch with their defences, noticing when there was avoidance, being able to approach feelings, whilst also being aware of dissociation and the fear of overwhelm. Some expressed low-level anxiety, and it was noted by participants that those with children found it much harder to allow feelings of despair.

*I guess my overriding feeling at the moment is fear … I feel it in my stomach … and there's grief for what's being lost … there's the fear of not being able to protect my children … I think that's where I really struggle …*

*I can't bear the thought of my children … (cries) … I don't know if it's a block that I have to believe that it's gonna be OK in whatever form that takes … I could easily fall into despairing, but I feel like I have a kind of faith in humans and the future …*

*I've really noticed as people talk that children come up a lot and I'm conscious that not having children plays a huge part in my thinking around it … I have a freedom to think and feel things that I suspect others don't …*

*Yeah, I think you're right there. I would say I don't feel despairing. It's almost like I can't … how could I? I think I have to hold the hope. It's not necessarily that we're going to find the solution but it's something about learning … the wisdom that will emerge … however it is … wiser … and also the 'inevitable' I'm not sure about.*

The conversation turned to solace.

*I mean gardening and the allotment definitely helps and I can still get into a place where despair and helplessness isn't constant but it's a hope that … well the planet's going to continue after humans have gone. I know that sounds really bleak but there's something in that – a hopefulness (laughs) … it sounds so weird saying it but maybe the human race doesn't deserve it …*

*I think there's a lot of solace to be found in that – I don't think it's bleak at all – to come back to our own inevitable death – there's a lot of peace to be found in the fact that life continues … and if you kind of zoom out enough there is a lot of gratitude to be found for existing in the first place (laughs) …*

*I too find the zoom out very comforting … at the same time I feel – as you said – that gratitude for having life in the first place, and for me what gives sense to the fear and the passion is that quintessential life force – the very gift that we have this life force …*

*I know that just holding the vitality of being alive … is absolutely essential … if we all disappear into despairing or not just getting together and talking about it or communicating … just living a life … we're lost.*

Within this part of the discussion there was a clear movement from expressions of despair through to gratitude, reminding me of Macy's despair and empowerment work (1983) and more recently the *Ecological Awareness Cycle* (Hawkins and Ryde, 2020) and Hoggett and Randall's 2018 research study showing climate activists' trajectory moving from crisis through to resolution and epiphany.

In discussing defences, the participants talked about knowing they were creating distractions as a defence, but some saw this as a strategy that allows them to cope, and that whilst some defence mechanisms are healthier than others, all of them are somehow allowing the organism to continue and survive. Blaming others was named as a psychological defence as well as guilt and self-blame. Some questioned whether art and workaholism were defences, and there was also some discussion around activism as a defence – pointing the finger at others so as not to feel powerless.

Turning towards what therapists might need in order to deal with these issues, I questioned whether they might need more information – data perhaps? This relates to colleagues' questions about the training curriculum, and considerations of future training input. Most felt that there was no need for more factual information:

*I would expect to have the same degree of knowledge about it as other topics that a client is likely to bring … if a client comes to me to talk about miscarriage for example I'm not an expert on miscarriage but I know enough to be with them, and I know enough to know that I'm ignorant as well and I can sit in my ignorance with that client and explore together. Does the topic of climate change require more from us, or does it require that same level of enough and knowing that there's more that I don't know?*

However, all participants said they *would* like to know more about how to work with it in clinical practice. They noted that they had no training input and questioned whether they needed separate 'Eco-training':

*I think there is such an absence of looking at relationship with place and land and the environment – not just in training but in the field of psychotherapy …*

*I went through the training wondering why all these theories about how I've become me are based on relationships with people when it's relationship with landscape that's built me … so putting that back in is a crucial step because that's the foundation for our feelings around the planet …*

*There's something about the focus on intimacy in therapy … if that's resolved everything will be resolved … including your relationship with the environment … and yet I'd have really liked my therapist to be brave enough … why was I pinned in my chair and why couldn't I say let's go outside?*

*Or even the boundary of the chair … like to get up. I'm gonna do it now! (stands up). It's like we're all just in our heads … this idea of animal body … I don't think it is a distraction, it's like we don't move …*

*Something about working with any kind of trauma … in order to do that we need to be able to sit with our own – we need to be able to hold that space, to go there with them … so more of this kind of thing – feeling our own feelings about this kind of stuff in relation with other people … how can we hold that space if we haven't practised holding it for ourselves and holding it with others?*

*I think maybe more connection to our own living bodies which is what connects us to other organisms. I think that is the doorway … that is how we have a relationship with the living world around us … and anything that gets us in here and from here into our living context has got to be helpful …*

*I resonate with what everyone has said … as someone who works outdoors and within the ecological system of the client, I think in understanding nature and self, the body is key to that … the senses … that animal self …*

## A Return to the Body

> Ecopsychotherapy recognises that we are embodied creatures, human animals. Our embodied self is not only our experience through our sensory worlds but through feeling and intuition also. Ecopsychotherapy recognises that all these creaturely aspects of being human need fostering alongside, and in relation to, the intellectual life of the human mind.
>
> (Rust, 2020)

The link suggested by the focus group participants between a more embodied awareness and a connection to the natural world is commonly understood and echoes my previous research into the phenomenon of 'body consciousness' (Staunton, 2008), explored through a series of body-led interviews, using techniques of 'free association through the body' (Boyesen and Boyesen, 1981, 1982) followed by semi-structured interviews to discuss the experience. Some of the participants in this study reported an enhanced connection to their environment, engaging them with visceral experiences through smell, visual and proprioceptive channels.

Environmental philosopher Paul Shepard speaks of 'the self with a permeable boundary … whose skin and behaviour are soft zones contacting the world instead of excluding it' (Roszak, 1995: 13). Some of my participants illustrated this – for example:

> *Horizons were more expansive … I felt more accessible to other people, to my environment … more open, expansive … it wasn't a different connection to the environment, but I stopped and gave it time*

> *I was acutely aware of smells … cut grass … lavender … I couldn't get enough of it*

> *I felt spatially aware … there was no edge between me and my environment*

For some, this connection to the environment was experienced as a spiritual awareness:

> *It made me acutely aware of other dimensions, and opened me up*

> *felt I had a more permeable membrane*

> *I was present in eternity … not limited by time and space*

Two participants had animal dreams following the interview:

> *White horses dancing around me … one horse came behind me and put his head – heavy and soft – on my shoulder …*

Abram tells us that 'ultimately, to acknowledge the life of the body, and to affirm our solidarity with this physical form, is to acknowledge our existence as one of earth's animals, and so to remember and rejuvenate the organic basis of our thoughts and our intelligence' (1997: 47).

My research findings serve to illustrate the body's part in countering the tendencies of modern culture to pull us out of a felt connection to ourselves and our environment. But more than this, they demonstrate how easily we can be drawn back *into* relationship with ourselves, our bodies and our environment, if we are listened to and listen to ourselves in an attuned manner. Over many years of teaching, I have introduced a *body-oriented* approach to trainees by inviting participants to write a letter to their bodies. What follows is almost always an apology, tears and regrets, the recognition of abuse and neglect. Following this, I ask them to invite the body's response, to write *from* the body (writing with the non-dominant hand assists this). The body almost always voices an immediate relief at being recognised, and forgiveness that is unexpected and perhaps unwarranted. 'I have been waiting for you to notice me'.

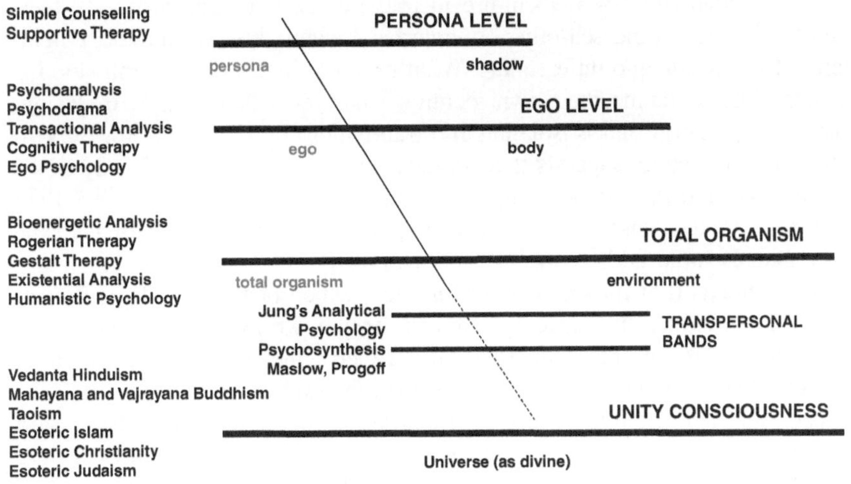


*Figure 13.1* Therapies and Levels of the Spectrum.

Reproduced from Ken Wilber, No Boundary: Eastern and Western Approaches to Personal Growth (Boston: Shambhala Publications, 2001), p. 14.

In *No Boundary* Ken Wilber asks, 'Do you feel you *are* your body, or do you feel you *have* a body?' He continues, 'Most individuals feel that they *have* a body, as if they owned or possessed it, much as they would a car, a house, or any other object' (2001: 5). Wilber discusses the *Perennial philosophy* – present in all major religions – which describe the expanded awareness that is possible when we do not draw the line between me/not-me, and it is possible to see our bodies, our environment and the entire Universe as part of who we are.

Figure 13.1 maps how the major schools of therapy address each of the levels of Consciousness as described in Ken Wilber's 'Spectrum of Consciousness.' He classifies five different levels of identity in terms of self/no self from Persona, to Ego, to Total Organism, to Transpersonal through to Unity Consciousness. At each level the sense of identity or 'I' is identified as belonging to the Self or outside of the Self.

The distinction of 'me/not me' is a process of identity formation that Donald Winnicott sees us attempting to resolve in early life, through a separation between self and other, seen as a developmental achievement. Is it necessary to first find a separate 'I' or is this a Western assumption? Hillman says this is the one core question for all psychology. Where do we 'make the cut' between me and not me? (1995: xviii).

I was fortunate to be present in a small group meeting in the early 1980s in *Plum Village*, where Buddhist Zen Master Thich Nhat Hanh presented the opposite truth. In Eastern cultures, the Self does not develop through separation but through *inclusion,* extending the sense of self to the family, the environment and the community, such that a wider identity is found. I am you and you are me – a Buddhist Truth.

In our divided and dis-eased culture, identity politics have separated us further, such that the me/not me, self/other boundary is drawn tighter around the 'I' in an attempt to delineate and differentiate. What began as inclusivity – permission for the individual to define their own identity – has too often become exclusion, causing a separation that is isolating and traumatic. Elif Shafak (2020), Turkish writer and intellectual, suggests that we have come to understand identity based on *exclusion*, and almost never upon 'multiple belongings' where, rather than attach to a limited sense of self, we might experience ourselves as citizens of the world, and, I would add, people of one earth.

Shafak posits that information technology, which began as a democratisation of knowledge, has endangered our deeper knowing; we have more information, but we are numb to what we hear. She says we are becoming tired, indifferent, desensitised. But is this psychic numbing a necessary defence against overwhelm? McKibben (1996) believes we have become 'information mediated' – we learn more about the world through information devices than through actual experience with the world. He says that the problem with virtual experience is that the *context* is missing. Our experience is no longer connected with our relationship to the world. McKibben says that people will come to view the real world as a commodity – like the experience they are buying in hyperreality. Ecopsychologists suggest that distancing or alienation between self and our natural environment is a 'repression of cosmic empathy; a psychic numbing we have labelled normal' (Roszak, 1995: 11).

We have become increasingly divorced from our lived experience, and our environment. Our senses are numbed. A loss of embodied knowing and relationship to the natural world increases the distance between subject and object, making it harder for us to grasp what is really happening in our environment, whilst body consciousness 'draws one into an immediate and direct contact with another person or with an object' (Stein, 1998: 12).

## Body and Soul

Over the last 20 years, the neuroscience revolution – much lauded by therapists of all stripes – is evidenced to endorse methodologies which 'use' the body as an avenue for healing trauma. The development of trauma-led research united psychoanalytic, cognitive and experiential approaches, giving rise to trauma-informed practices within the NHS, and fostering acceptance of a more integrated approach (Levine, 2008; Van der Kolk, 2014; Ogden, 2006, 2022). These body-based trauma practices will be clinically important as the climate crisis deepens.

However, in my research I have used the term *Body Consciousness* (2008) because it suggests something closer to Soul psychology than Ego psychology. In practice this means to *experience* and *deepen* into, and accept *what is*, rather than hold the intention to change. Jungian Robert Stein notes: 'My individual experience of the movements of a living pulse, a vital, ever-changing energy within my body, is what I understand as soul' (1998: 17). It is the quality of attention to the experience that returns us to our senses. It requires us to slow down. The mind processes thoughts at speed, whilst the body assimilates our experiences gradually over time. Psychotherapies that embrace a phenomenological approach follow the moment-to-moment experience that is present within the client, the therapist and the 'between', thereby facilitating a *somatic consciousness.* It is this translation from the somatic unconscious to a somatic consciousness that offers us the potential to engage with the natural world and our animal selves. This is not to propose the body as saviour or sanctuary, just as Nature is not simply a green prescription. Connecting with the psyche-soma level is a *descent*, where our basic wounds – individual *and* collective – are opened.

> Jung writes of the descent to the plant level as 'the downward way, the yin way … (to) earth, the darkness of humanity'.
>
> (Brinton Perera, 1981: 58)

Connection with the natural world is not a means to an end, something we must aspire to as an ego-ideal, rather it arises naturally when we connect more deeply with our own nature. Furthermore, when we experience a 'more permeable boundary' we are less separate and less defended against the 'cosmic empathy' that Roszak speaks of. Jung affirmed 'The body's carbon is simply carbon. Hence "at bottom", the body is simply "world"' (Jung, 1970: 291).

## Conclusion

> Once upon a time all psychologies were 'ecopsychologies'.
>
> (Roszak, 1992: 14)

It is well past time to effect the 'greening of psychology' where Psyche meets Gaia (Roszak, 1995: 16). The realisation of what we have done to our earth brings a complex web of emotions – our collective guilt and regret threaten to overwhelm us as we take in the truth of our situation. We must grieve our losses. Robert Sardello has described grieving as 'an awakening of soul consciousness in the body' (1995: 99). Dwelling in grief we are open to the Universe. When we rest in a more embodied consciousness we can embrace dualisms more easily, accepting what is now and what is to come.

The notion that therapy could take place as effectively online as in-person became a popular belief amongst therapists and regulators during the COVID-19 pandemic and many have continued in online practice (American Psychological Association, 2021). Whilst there is no doubt that it has been extremely valuable in increasing access to psychotherapy, I would argue that our embodied presence is key to co-regulation in therapeutic work. Shafak comments that as the world becomes more polarised, we see a decline in empathy in all cultures across the world. Perhaps this is also a result of our increased reliance on digital communication and distancing from each other. Empathy is an inherently embodied experience requiring energetic attunement – a process known as *somatic resonance*. Whilst somatically trained psychotherapists have developed skills in this area, we are all tasked to neither limit nor exclude somatic reality in our relationships.

The challenge in practice is not only to invite the external world of our faltering culture into the therapeutic space to assist psychic integration, but to move the therapeutic space out into the world. Moving forward with regulation, the next step is for all regulatory bodies to follow the UKCP Humanistic and Integrative College and include *environmental breakdown* as our current context of practice.

The emergence of many forms of Ecotherapy and the increased interest in Ecopsychology, along with the development of practices involving the body, suggest that the focus in our profession is shifting, redressing the balance from *ego-centricity* to *eco-centricity*. I propose that *all* therapies, regardless of their orientation, must 'turn towards the body' as a necessary step in 'coming to our senses'.

## Notes

1 United Kingdom Council for Psychotherapy.
2 www.psychotherapy.org.uk/media/e4aijiuo/hipc-specific-standards-of-education-and-training.pdf
3 The group consisted of four graduate therapists, one psychotherapy trainee and two foundation-level students who had an opportunity to see the transcript and agreed to extracts being published anonymously.

## References

Abram, D. (1997) *The Spell of the Sensuous*. New York: Vintage Books.

APA 'Online Therapy is here to stay.' Apa.org [Accessed 8 May 2023] www.apa.org/monitor/2021/01/trends-online-therapy

Boyesen, G. and Boyesen, M. L. (1981, 1982) *Collected Papers of Biodynamic Psychology*, Volumes 1 and 2, Biodynamic Psychology Publications: London.

Brinton Perera, S. (1981) *Descent to the Goddess*. Toronto. Inner City Books.

Hawkins, P. and Ryde, J. (2020) *Integrative Psychotherapy in Theory and Practice*. London: Jessica Kingsley.

Hillman, J. (1995) A Psyche the Size of the Earth: A Psychological Foreword. In: T. Roszak, M. Gomes and A. Kenner (eds) *Ecopsychology: Restoring the Earth, Healing the Mind*. Berkeley, CA: Sierra Club Books.

Hoggett, P. and Randall, R. (2018) Engaging with Climate Change: Comparing the Cultures of Science and Activism. *Environmental Values*, 27: 223–243.

Jung, C. G. (1970) Collected Works Volume 9i *The Archetypes of the Collective Unconscious* (2nd edition) London: Routledge.

Levine, P. (2008) *Healing Trauma*. Boulder: Sounds True, Inc.

Macy, J. (1983) *Despair and Personal Power in the Nuclear Age*. Philadelphia: New Society Publishers.

McKibben, B. (1996) Out There in the Middle of the Buzz. Forbes, ASAP Magazine, pp. 107–129.

Ogden, P. (2015) *Trauma and the Body*. New York: W. W. Norton & Co.
Roszak, T. (1992) *The Voice of the Earth*. New York: Simon & Schuster.
Roszak, T., Kanner, A. and Gomes, M. (eds) (1995) *Ecopsychology: Restoring the Earth, Healing the Mind*. San Francisco, CA: Sierra Club.
Rust, M. J. (2020) *Towards an Ecotherapy*. London: Confer Books.
Sardello, R. (1995) *Love and the Soul – Creating a Future for Earth*. NY: Harper Collins.
Shafak, E. (2020) *How to Stay Sane in an Age of Division*. London: Profile Books.
Staunton, T. (2008) Explorations in Body Consciousness. *Self & Society*. 35, Association for Humanistic Psychology in Britain.
Stein, R. (1998) *The Betrayal of the Soul in Psychotherapy*. Thompson, CT: Spring Publications.
Van der Kolk, B. (2014) *The Body Keeps the Score*. New York: Viking.
Wilber, K. (2001) *No Boundary*. Boston: Shambala.

Voice 9

# I want to fly

*Frankie (pseudonym)*

A few years ago, a 12-year-old called Frankie was asked how they felt about the climate crisis. That was me, and I am now 18 years old. Back then my response was 'Arghhhh' or something like that, not really an answer but it was the best way I knew of expressing how I felt. When talking about the climate crisis, sometimes the sheer weight of the topic can be overwhelming, and I think that 12-year-old me expressed that in a way that some people can relate to. But now I have a different outlook, back then I said, 'it's hard to explain', and to some extent that is still true, but since then I've found a way of dealing with those emotions in what is for me a productive way, instead of just panicking.

In the past few years, I have expressed a growing interest in Aviation, which I'm sure you know is not exactly the most sustainable industry, but despite that my heart is set on becoming a commercial pilot. Has the climate crisis deterred me from choosing this career? No. Instead of worrying about how my dream will affect the planet I chose to educate myself on how *it* can change to a sustainable future of Aviation. I always knew that there was technology that could make Aviation sustainable, but I didn't know how advanced it was and how effective it would be. So, I spent my time researching; I wanted to find out if the industry I want to be a part of can ever be sustainable. Luckily, I found out some promising information which proved to me that I should feel no guilt in choosing that career. Did you know that Airbus aircraft are capable of flying using 50% Sustainable Aviation Fuel? I didn't. Did you also know that Airbus have made it their goal to make all of their commercial aircraft compatible with 100% Sustainable Aviation Fuel by 2030? Neither did I. Of course, this technology will take time to be applied globally, and it will require significant investment from the Government, but it is there, and it's only a matter of time before it's used for all commercial flights.

So that's what I did, I gained knowledge, and it has enabled me to feel comfortable that I can be part of the solution, not part of the problem. Yes, I am joining an industry which contributes to the climate crisis, but if not me then someone else would, and it's my dream, so I'm going to do it anyway.

DOI: 10.4324/9781003436096-26

I am aware of the issues, and I am hopeful that in the near future there will be significant improvements.

So whenever I start to feel overwhelmed when thinking about the future, I just remind myself of the facts and it helps me to see clearly. Back then Frankie may have said 'Arghhhh', but what good does it do to panic? Now Frankie says relax, because I know what's happening to the planet, and I know how I will react to it.

Frankie, UK

Section Four

# The Ecological Self

Chapter 14

# The Zone of Encounter in Therapy and Why It Matters Now

*Kelvin Hall*

## Introduction

In this chapter I explore how the 'Zone of Encounter' can restore a sense of being 'at home' on the Earth, offering relief from the stress of climate-anxiety while complementing and invigorating activism. It could instil a vision of a new ecological culture. To describe the 'Zone' I will draw on the work of Abram (1996), Jensen (2002) and Bernstein (2005) and offer an exploration of the ways in which it enhances therapy by fostering 'present moment' experience. I also draw on concepts from Kohut (2013) and Stern (2004) and offer a number of hypotheses to support its relevance in the context of climate emergency.

## Meeting in the Zone

After practising as a therapist for 20-odd years, in about 2008 my work began to include a new focus on what I have come to call the 'Zone of Encounter'. This is a realm of experience in which human and other-than-human life converse intimately with each other, illustrated below. It can be part of both the content and the process of therapy. This new focus came about as I reconsidered moments in my own life when the eloquence of other creatures (or even trees) had been transformative; for instance, a horse becoming calm and co-operative as soon as I deepened and steadied my own breathing. I found comparable incidents in other biographical narratives – some published, some communicated personally to me when I began to research this subject. I perceive a crucial connection between the emergence of this facet of therapy practice and the growing urgency of the climate emergency, explored in the latter part of this chapter.[1]

The phenomenon of meeting and exchange between human and other-than-human beings, in which there seems to be lucid and mutual understanding which frequently defies rational pre-conceptions is examined in the work of Jensen (2002), Bernstein (2005) and Abram (1996). They also substantiate the idea that an unconscious exchange with the living world in

DOI: 10.4324/9781003436096-28

all its variety is happening much of the time. Abram writes of 'this silent and wordless dance already going on – this improvised duet between my animal body and the fluid breathing landscape that it inhabits' which he finds 'whenever I still the persistent chatter of words within my head' (Abram, 1996: 52–53). Their work has been greatly consolidated in recent years by writers like Suzan Simard (2021) on the relational attributes of trees, or Eva Meijer (2020) on animal languages, who urges a reconsideration of what actually constitutes language. My focus here is not so much the objective truth of what happens (e.g. how much do animal and human really understand each other?), but what the dialogic experience entails for the human's state of being and the quality of their relationship with their ecosystem. A growing number of volumes testify to the ways the human can be healed by immersion in the natural world. For instance, Stuart-Smith (2020) describes the trauma-healing benefits of horticulture – of active participation in cycles of decay and renewal which are vividly real to the senses, to sight, sound, smell, touch, and which both challenge and validate the human's role in these cycles. I submit that the Zone of Encounter represents a particularly intense and relational version of this immersion.

An example of what I mean by the Zone is provided by a series of incidents which a client narrated during sessions. In early adulthood, she had suffered multiple bereavements and was 'drowning in grief'. Then, while on a walk, she came upon a group of several moles in the grass. They showed no fear and one of them allowed her to stroke it. Her dog, rather than try to chase the moles, sat down quietly and watched. Soon after, while she sat in her garden, a young fox came through the garden hedge, sat with her, exchanging eye contact for a prolonged period. The fox then stayed waiting while she went indoors to find a lump of cheese, received the cheese from her hand, and later, after a further stay, departed. These encounters were decisive in her regaining a sense of belonging in the world. When the stories arose as part of the *content* of sessions, my main role was to validate them through the quality of my attention.

Later, however, a crucial part of our *process* occurred in the Zone. We were meeting in a secluded meadow, where I often worked. The woman was once again feeling discouraged and alone. This was triggered by some critical comments from a friend and then from a partner, which made her feel deeply isolated and inclined towards withdrawal from life and human contact. These arose against the background of a new bereavement – the loss of a beloved dog. A parallel process then arose when I made some comments which she heard as critical, and she came close to leaving the therapy. I offered the perception that, rather than rejections, such comments can be seen as attempts to achieve a fuller meeting and can be responded to as such. She realised that it was possible to make a choice and was on the verge of deciding to recommit to the therapy instead of leaving it. At exactly that moment, a very large fox entered the meadow, stopped about five yards from where we sat, and stared at us, before eventually moving on. The effect of this was profound. Particularly in the light of her previous fox experience, it was as if she received a message of welcoming and encouraging, and affirmation of her belonging. Following this, she did indeed re-engage with therapy and her personal relationships.

## Links to Theory

I find Kohut's theory of *selfobject function* (2013) helps me to understand what is happening in these encounters. We can regard both fox incidents as examples of *mirroring* (ibid.: 197), a deep affirmation of the being of the self by the other. Often such mirroring is experienced quite literally in the way one is looked at by the other, the message implicit in their eyes. Such moments enable the self to come more fully into existence. When visited by the fox in early adulthood, the woman described herself as being in a state of dissolution at that time, her losses *'breaking her soul and her body … as if she had no place on the earth … no one … who needed her or recognized her'.* The animal encounters were decisive in enabling her to live again.

The exceptionally powerful impact of such incidents has a number of aspects. The animal is seen as freer and less conditioned by social role or the constraints of civilised behaviour, for example the therapist who has a professional obligation to be 'supportive'. Like our early life contact with others, these exchanges are essentially non-verbal. Stern (2004) explores the quality of exchanges between humans, which are wordless and therefore very immediate, such that their awareness is entirely in the present moment. These non-verbal exchanges are also moments of what he calls *Implicit Relational Knowing* (ibid.: 242). Immersed in the here-and-now encounter, such habitual distractions as anxieties about the future and disappointments about the past fade away. He further asserts that the use of language is closely linked with consciousness and sense of self – we become conscious as we begin to place our experiences in a narrative which he calls *lived story*. But as soon as we do that, some of the immediacy is lost. In the transition between immediate experience – *the present moment* – and *lived story* – the human creates a narrative of the immediate and thus becomes conscious of it. I speculate that in moments of sustained dialogue with other-than-human life we hover continuously between the *non-conscious* and *lived story.* Experience in this area can seem pristine. We enter more fully into the present moment, less engaged in anxiety about the future or the past. This is often experienced as profoundly refreshing.

One of the things it restores to us – to draw on Stern again – is the pre-verbal world we lost with the advent of language. In describing the pre-verbal perception of the infant, he writes 'It makes a rich, multimodal sensory-feeling world'. He contrasts this with the world as experienced with the advent of language: 'The original world is gone. (Ibid: 144) … . Something is gained and something is lost when experience is put into words. The loss is wholeness, felt truth, richness and honesty'.

I suggest that the non-verbal exchanges I have narrated restore something of the freshness of the pre-verbal world. Indeed, the woman visited by the fox wrote that '*every detail of the little fox delighted me and astonished me as if my eyes were seeing more than they could usually see, every hair on the fox's fur and every glistening speck of light on every hair*'.

In other accounts of such immediacy, there is often reference to having regained something which had been lost. Sometimes this is explained as a loss that took place in childhood, of a sense of intimacy with the other-than-human world. One woman recounted her earliest memory; she was lifting earthworms to her lips, immersed in the sensual thrill of the contact. But she also remembers the anxious cry of 'Stop her!' yelled from one parent to the other, who was at that moment in closer proximity to the child. She then recalls a childhood in which the natural world was kept distant and was regarded as a threat to cleanliness and order. Jane Goodall (May, 2016) similarly recalls being discovered as a toddler, in bed with a handful of earthworms, and sees this as a precursor to a lifetime's fellowship with animals, but one in which she had to argue resolutely for the validity of their emotional life, and therefore their soulful connection with us. Another wrote:

> My childhood days were spent with beloved family dogs, cat, goat, horses, and the ducks. I watched for hours the minnows in the pond, the resident swans in the meadow, and the field mice. My childhood was embedded in the natural world.
>
> (Moultrie, 2017)

Tudor (2014) graphically describes her experience of regaining her childhood intimacy with nature as *homecoming*. While for some this home-coming can entail finding an actual geographical place where one has a sense of belonging, for others it is more definitely about entering a state of mind. This is amplified and echoed in the words of individuals who have achieved a dialogic partnership with wild creatures. When conservationist Lawrence Anthony (2010: 85) was struggling to earn the co-operation of a herd of traumatised and much-feared African Elephants, he had a break-through moment in a fraught confrontation with the matriarch of the herd. When he persistently held in mind his intention of conserving the life of the herd, she softened and touched him with her trunk. In trying to describe the mood that this stimulated in him he uses phrases such as 'I had never felt safer', and 'I was in a bubble of well-being'.

For many of us, if not all, there may be an inherited ancestral loss of intimacy with the Earth, which can be traced to a specific social upheaval, or it can be seen as more universal. The several waves of Enclosures in Britain, when common lands were fenced off to enable landlords to graze large flocks of sheep, deprived country dwellers of grazing, foraging, firewood and wandering rights; they therefore also lost the personal engagement with their surroundings. Likewise, the Highland Clearances of the 18th century (in which – according to some accounts – whole populations were forcibly removed to towns or to America) left an imprint of anguish and bitterness. The narratives of dislocation and loss are many and varied. Nicolson (2008: 218) chronicles the collapse of a period of rural equilibrium with the coming of the English Civil War. MacDonald (2014: 103–104) alludes to the search for 'solace and safety' which motivated the growth of rambling in the 1930s, as people 'walked backwards in time'. Many of us must carry in our family history events like these; variations on the theme of lost connection. They may indeed have been handed down to subsequent generations in the guise of unspecified anger, grief and longing. If the 19th-century Irish Famine and emigration are part of one's heritage, if one has African or Asian roots reaching back into times of slavery or colonialism, the inheritance of dislocation may be even more acute. For the refugees and migrants of the present day, of course, the reality of dislocation is all too literal and immediate.

The narrative of a collective loss of dialogic intimacy with nature informs the work of all three of the writers mentioned earlier – Jensen, Bernstein and Abram. It receives some corroboration from anthropologists such as Nelson (1989), who describes the soulful and reciprocal attitude to trees among indigenous people in Alaska. Other commentators categorically link this loss to the introduction of agriculture (Glendinning, 1994) while Taylor (2008) traces it to the onset of a profoundly disruptive climate event at a time when the transition from hunter-gatherer culture to agriculture was taking place. In his account, a traumatic drought that afflicted a large area of the globe undermined the trust and reverence for nature which had characterised the human psyche until that time. Thereafter human society became more territorial, aggressive and hierarchical.

In many mythologies across the world the image of a lost realm occurs – Eden, the Golden Age of Greek Mythology, and their equivalents in Native American and other cultures. This re-appears in other forms throughout the history of literature – for instance in the portrayal of the Forest of Arden in Shakespeare's *As You Like It*, whose inhabitants 'fleet the time carelessly, as they did in the golden world' (Act 1 scene 1). In more recent works, Lewis's Narnia and Tolkien's Valinor maintain the tradition. I am well aware that the factual accuracy of any of the historical narratives can be contested. But the proliferation of such stories of loss implies to me that for very many people they make sense. Even if it is not historic fact, the myth so pervades the human psyche that encounters in the Zone *feel* like a restitution and are therefore deeply comforting. The different versions of loss become interwoven:

Ruth Tudor (2014: 21) writes: 'This feeling of being very far from home – from who we deeply are – is a collective trauma. My personal loss of home at eleven resonates with rupture (sic) that is shared across the culture in which I live'.

Furthermore, those past losses echo through the current sense of dislocation which is widespread even among those living a relatively stable life in Western Europe. Now it also resonates with current and forthcoming loss due to climate change. This faces us with a new uprooting on a massive scale. It is likely we are all going to be refugees; even if we stay in the same place, that place will not be the same. The message of the climate emergency is that we are all losing our homes, that we have broken the bond with other life which maintained a balance. At some juncture, we stopped listening to the Earth. However, time spent in the Zone resumes the conversation, restoring the bond which the climate emergency tells had been broken. Just to experience that, even if briefly and even if it does not change the final outcome, can be profoundly refreshing.

The phenomenon of eco-anxiety is not new but has increasingly entered the discourse of both the mass media and the therapy profession. In about 2009 I had a client who told me that the onset of depression in his life was triggered by the information delivered by his teacher in secondary school about the multiple threats to the environment posed by human activity. He told me it was his firm belief that much current anxiety and depression in the wider population actually had that origin. In a 2020 post on a climate psychology network, a correspondent wrote that he had spent time as a client with two therapists, hoping for help with the distress he had felt deeply as the implications of the climate crisis had sunk in. But neither of them seemed to him to recognise that the distress was justified, and this increased his despair.

Other therapists, though, have come to the firm conclusion that alarm is a sane and realistic response to the signals we receive from the natural world and the information we receive from the media. With another of my more recent clients, this has been one of the most prominent and recurrent threads in the work. She fears for herself and her children in an era of extreme weather events, the breakdown of society and loss of food supplies. We have examined the history of trauma and dread in her biography and scrutinised the way these may colour her current reactions. Nevertheless, she is also adamant that she receives consolation from my support for her perception of the ecological dangers as real likelihoods. Her conviction motivates her to active participation in Extinction Rebellion campaigns and other forms of social initiative, such as setting up a network prioritising local community co-operation and mutual support. This approach to therapy means that I have to tolerate my own fears being present in the room when we work, as well as hers. At the same time, in such cases, I also look for opportunities to approach the Zone of Encounter. Alongside these other resources, the Zone offers a world restored and the possibility of a world that might be in the future.

If we and our world are going up in flames or drowning in floodwaters, I have the choice of being *with* it, rather than still behaving as if separate from it, *whether or not that makes a change to the final outcome.* Of course, there may be something absurd about extolling the wonder of mutuality, when the other party – and of course ourselves – are on the verge of calamity or extinction. However, this also may take us towards a spiritual sphere of relational connection, beyond the usual perceptions of time, space and what is gainful. This dimension is conveyed for me in the extraordinary story which a colleague narrated to me a few years ago. Whether it is factually true, which I've been unable to verify, is less important than what it evokes. Although the events of the story are confined to the human sphere, it seems to me that parallels can be drawn with the relationship between human and other in a time of extinctions. A nun who was a prisoner in a Nazi concentration camp witnessed a column of other inmates being herded to the gas chambers. Among them was a child, distraught and alone. The nun was not due for extermination, but she chose to join the column, take the child by the hand and comfort them, passing into the gas chamber with them. In a way this is an absurd gesture that didn't save either of them. In another way this was a gesture of the deepest compassion and soulfulness.

I offer another relevant image from my own practice, which arose from a session that contrasted greatly with the bucolic quality of some equine and outdoor therapy sessions. A woman once contacted me wishing to experience equine-assisted therapy. I arranged to meet with her, but on the day of our first meeting there were fierce winds, and a dust cloud, blown up from the Sahara, turned the sun over Western England deep red. It looked apocalyptic and people afterwards referred half-jokingly to that day as 'the End of the World'. We persisted with our meeting and had an initial discussion in a secluded meadow where I often work, before taking a short walk to the next field, where the horses were. But the first one we came to was lying on the ground with his teeth bared, evidently in considerable discomfort. In fact, as became clear later, the horse was dying. It was suffering, I later learned, from severe colic on top of a stroke some days previously. Emergency care for the horse became the top priority. We spent time sending messages to the horse's carers and vet; but we also offered the horse the full measure of our company and attention. When the time came for the session to end, I suggested that the client left while I waited for the vet (I later notified the client of the horse's death). Hypothetically then, this could have been a disastrous first session. But it was the beginning of a year's intense work on connection with animals, underpinned by the realities of death and the limitations of time. It seemed to me that the urgency of our attempts to respond and accompany the dying horse on that first day gave the rest of our work a rare intensity.

## Conclusion

If humans maintain the ability to enter the Zone, this may preserve something of value on which to eventually build a wiser future for our species. But even if that is far too optimistic, entering the Zone can retrieve present time from total submergence in fear, anxiety and the losses of past or future. It can offer an image of restoration which is deeply heartening *whether or not* it ever actually changes the future. In cases where present time itself consists of trauma, the existential decision to remain in the Zone carries even more profound implications, and goes, I think, beyond the realm which words can adequately describe.

## Note

1 Permissions received from all parties for material quoted.

## References

Abram, D. (1996) *The Spell of the Sensuous*. New York: Vintage.
Anthony, L. (2010) *The Elephant Whisperer*. London: Pan Books.
Bernstein, J. (2005) *Living in the Borderland*. Hove: Routledge.
Glendinning, C. (1994) *My Name is Chellis and I'm in Recovery from Western Civilization*. Boston: Shambala.
Goodall, J. (2016) *Private Passions*, BBC Radio 3, May 2016.
Jensen, D. (2000) *A Language Older than Words*. London: Souvenir Press.
Jones, L. (2020) *Losing Eden: Why Our Minds Need the Wild*. London: Allen Lane.
Kohut, H. (2009) *The Analysis of the Self: The Systematic Approach to the Psychoanalytic Treatment of Narcissistic Personality Disorders*. Chicago: University of Chicago Press.
Kohut, H. (2013) *How Does Analysis Cure?* Chicago: University of Chicago Press.
Macdonald, H. (2014) *H is for Hawk*. London: Jonathan Cape.
Meijer, E. (2020) *Animal Languages*. London: John Murray.
Mitchell, E. (2023) *The Wild Remedy: How Nature Heals Us*. London: Michael O'Mara Books.
Moultrie, D. (2017) Relational Learning from the Natural World. *Unpublished MA dissertation, Bath Centre for Psychotherapy and Counselling*.
Nelson, R. (1989) *The Island Within*. New York: Random House.
Nicolson, A. (2008) *Arcadia: The Dream of Perfection in Renaissance England*. London: Harper.
Simard, S. (2021) *Finding the Mother Tree: Uncovering the Wisdom and Intelligence of the Forest*. London: Allen Lane.
Stern, D. (1995) *The Interpersonal World of the Infant*. New York: Basic Books.
Stern, D. (2004) *The Present Moment in Psychotherapy and Everyday Life*. New York: W. W. Norton and Co.
Stuart-Smith, S. (2020) *The Well-Gardened Mind*. London: William Collins.
Taylor, S. (2005) *The Fall: The Evidence for a Golden Age, 6000 years of Insanity and the Dawning of a New Era*. Winchester: O Books.
Tudor, R. (2014) Becoming More Human in a Thickening World. *Unpublished MA dissertation, Bath Centre for Psychotherapy and Counselling*.

Chapter 15

# Rewilding Therapy

*Nick Totton*

Over 20 years ago, I began, along with others, to think about relating therapeutic and ecological ways of thinking, feeling and being. The tendency was – and very often still is – to simply bolt therapeutic ideas onto ecopsychology, or ecopsychological ideas onto therapy, without really bringing the two together into a new whole; for example, trying to preserve the conventional therapeutic frame while working outdoors, which feels a bit like dragging a big iron bed frame on one's back through hedgerows and ditches – inappropriate and hugely uncomfortable! It seems to me that a genuine synthesis requires *both* traditions to be challenged and changed. The here-and-now reality of climate and environmental collapse, brought into the therapy room every day unless the therapist rigorously excludes it, makes this an increasingly urgent priority.

For therapy to think and feel ecologically, it must change some of its core attitudes: for example its extreme individualism and its exclusive obsession with early family relationships. Many of the changes are needed because they involve an ecosystemic viewpoint which is central to many indigenous cultures, and also help decolonise therapy. Paradoxically, however, many elements of an ecological practice are already present in the field, often without being perceived as such. What I have therefore tried to do, for example in my book *Wild Therapy* (Totton, 2021), is to identify these implicit elements and assemble them in a coherent way, describing a way of doing therapy which is both new and always already present. A therapy that recognises that humans don't stand alone in the universe but are profoundly connected with and dependent on other species and entities with whom we share this earth; and also recognises that skilful living stems from a capacity for spontaneity and yielding to what is, rather than from a struggle to exert control over self and others.

The emphasis on spontaneity is as much part of ecological awareness as the emphasis on connectedness. Both ideas follow from systems theory, which sees the world as a set of complex, self-organising, adaptive systems, where nothing *causes* anything else in a linear sense, but everything mutually *responds* to everything else, in a way that parallels the Buddhist concept of

DOI: 10.4324/9781003436096-29

*paticca samuppada*, 'dependent co-arising' (Macy, 1991, 1995). Hence trying to separate oneself from the world and exercise control over it is ultimately self-defeating – as we are seeing with the current ecological crisis.

Therapy has often stood against the dominant cultural injunction 'Be in control of yourself and your environment'. It has tried to help people tolerate the anxiety of not being in control – of our feelings, our thoughts, our body, our future. There has always been a struggle over this issue, however: new forms of therapy constantly arise which claim: 'You can be in control after all – you can regulate your neurology, calm your emotions, direct your life'. And currently the dominant social structure is saying that therapy itself must be controlled, brought within the field of surveillance, monitoring, regulation and safety.

This parallels humanity's response to crisis. As the world becomes increasingly frightening, it seems increasingly necessary to pretend that security can be achieved through ever-greater monitoring, surveillance, barrier-making and censorship; and this process in turn ratchets up our fear and insecurity. This is the path our society seems to be taking; and equally, of course, it is an internal psychological process, embodying exactly the anxieties that therapy arose to address.

I developed Wild Therapy as a conscious intervention in this destructive, spiralling effort to control the uncontrollable. As a theory and as a practice, it points to the ever-present tension in human culture between the wild and the domesticated. In the move from hunter-gatherer society to agriculture, human beings tried to gain control over the world, over each other, and over the other-than-human and more-than-human. In doing this we split ourselves off from the world – it became, in fact, our 'environment', rather than the whole of which we are an integral part. In Ursula Le Guin's resonant phrase (1998: 366–367), we learnt to live 'outside the world'.

By trying to control the world, we have made it *other*, and therefore dangerous and frightening. The more we seek control, the closer we seem to get to it, the further our goal recedes. As the *Tao te Ching* tells us, the more we try to control things the further out of balance we push them: 'Do you think you can take over the universe and improve it? I do not believe it can be done. The world is ruled by letting things take their course. It cannot be ruled by interfering' (Feng and English, 1972: sections 29 and 48).

Therapy is, I believe, by nature wild; but a lot of it at the moment is rather tame. Wild Therapy is intended to help shift the balance back towards wildness, by showing how therapy can connect with ecological thinking, and hence with the mutual co-creation of all beings. When we think ecosystemically, we see each species, each being, not as an isolated monad, a sort of old-fashioned billiard ball atom interacting with other billiard balls by knocking into them – but as inherently and profoundly linked with every other species or being. We develop a sense of the endless complexity of existence; and realise that wildness, a state where things are allowed to happen of their own accord, is far more deeply complex than domesticated civilisation, just as a jungle – or even a piece of wasteland – is more complex than a garden.

Although the path of control is becoming increasingly emphasised in society, it has been with us since the Neolithic development of domestication. This was not just about humans domesticating other species; we also and above all domesticated *ourselves*. How far can we reasonably hope to go in rewilding ourselves, given that the sustainable forager population of the earth in the Palaeolithic era was perhaps around one person per square mile? In a literal sense, we clearly cannot go very far at all. A reduction of humanity to Palaeolithic population levels – which a few people are understandably desperate enough to hope for – will only happen through catastrophe; and only in this way could a few surviving humans live wild in the literal sense.

In *Wild Therapy* (Totton, 2021) and other writings I explore the possibility of developing a wildness that is less literal, but perhaps nonetheless real and important: a reconnection with what I describe as 'Wild Mind', which is necessarily at the same time a reconnection with the world and with the other beings which inhabit it – and a mode of being present in *all* cultures, but suppressed for many years in our own, and emerging only in marginal expressions. Wild Mind refers to a state of awareness in which humans will not want, or be prepared, to damage the world for our own short-term comfort and convenience. Reinstating it will involve thinking, feeling and living in very different ways.

Wild Therapy offers a context for all this through connecting the attitudes of forager cultures with contemporary Western understandings of consciousness; it suggests how this can be expressed through a therapeutic approach which is not newly invented but brings together a wide range of already existing ideas and practices. Some features I have proposed for Wild Therapy are summarised below

- It recognises the interdependence of everything that exists
- It is through-and-through relational
- It identifies the role of the other-than-human and more-than-human in the therapeutic process
- It supports, protects and defends liminality
- It celebrates embodiment as a central aspect of our existence
- It welcomes the spontaneous and the unknown, trusting what arises of its own accord
- It seeks to transform fear-based defensive practice into undefensive, contact-based, adventurous practice

These are discussed further in Totton (2021: 212). I would say that all of these features are present (though unevenly distributed!) in the therapy field already; but I think it is important to foreground and connect them in this way, and to show how they represent an ecological way of understanding. Wild Therapy begins by taking therapy outdoors to meet the other-than-human and more-than-human – the one-year training (for which see http://wild-therapy.co.uk/) is largely about taking therapy into the wild; but it is crucial to then take the wild back into the therapy room – in fact, to open oneself to the reality that *wildness has always been there.* Once we open our eyes and ears, we find that the animal and plant peoples constantly appear in the therapy conversation; that our clients are continually referring to their loss of connection with the world, and their fears for its survival. We plumb the depths of our own need to impose control over what happens in the practice room.

So Wild Therapy looks in one direction towards therapeutic practice; and in another direction towards the 'deep background' of human nature and human history. It also looks in a third direction and asks *how therapy can contribute to changing our relationship with the ecosystem of which we are part.* To avoid environmental catastrophe, the world needs a change of heart, a fundamental shift of consciousness and behaviour which is perhaps only slightly less hard to imagine than a return to hunter-gatherer lifestyle. However, it is necessary; without it, a literal return to hunter-gatherer life (and population levels) may be the *best* we can hope for.

I believe psychotherapy and counselling have an important role to play in supporting and facilitating Wild Mind, or ecological consciousness. Changes of heart are what therapy specialises in; and ever since it began, therapy has been trying to help the world change its heart, by offering it collective as well as individual therapy. The only problem, as Andrew Samuels points out, is that like many individual clients who at first seem enthusiastic, 'the world has not shown up for its first session. The world is ambivalent about its therapy, suspicious of its political therapists, reluctant to be a patient' (Samuels, 1993: 30). A suspicion and reluctance which we can also see in people's reaction to information about climate and environmental crisis.

This isn't hard to understand. Most of us are deeply traumatised, acutely or chronically, personally and/or by inheritance, and live in a society that as a whole is also traumatised; trauma gives rise to dissociation and denial. I trace this trauma to its origins in what I call the 'Neolithic bargain', when we exchanged the freedom and wellbeing of a hunter-gatherer lifestyle for the benefit of protection from other humans, but at the price of the increasingly damaged attachment which goes along with urban existence in patriarchal societies.

The structures of domination that this created now reach very deeply into our psyches and bodies; in viral fashion, they seek to take over and control every new and hopeful social formation that arises. As we are currently experiencing, they make it enormously hard for us to free our attention to deal with the environmental crises we face. In advanced capitalist society, nearly all of us are on the edge of being unable to cope, unable to do what we have to do and process what we have to process while also handling our internal emotional states. And a further level of this is *cultural* overwhelm, the result of many generations of damage through war, famine, disease and abuse.

Individuals therefore seek to protect a fragile bubble of personal reality that makes their life bearable. Some key elements of the bubble are fun, freedom, status-based identity and, most fundamentally, relaxation. People who talk about environmental catastrophe appear to threaten all of these elements, which in many of their most common forms – consumption, travel, entertainment – require high carbon consumption.

Such information seems above all to threaten *relaxation*: the human need for downtime, empty mental space, periods when we are not anxious and planning for survival. Even if we can only achieve relaxation through getting drunk and watching TV, it is still deeply precious, and we will protect it at all costs. Hence for large numbers of people it is not climate change *itself* which appears as a danger, but rather, *news* of climate change, which might break into their fragile bubble of emotional survival.

Until we are willing and able to tolerate the feelings that information about environmental crises set off in us – feelings of fear, grief, rage, despair – it will be very difficult for us to absorb that information, and therefore to act on it. Therapists know a thing or two about these feelings; we know, for instance, that the first thing to do when faced with overwhelm in a therapeutic situation is to point out to the person that this is what is going on: 'It's all a bit much, isn't it?' 'It's hard for you to take things in just now.' Just on its own, this helps people contact reality and find some solid ground. Then we can help build a sense of safety which will allow them to access their embodied emotions.

This approach can be carried over to collective overwhelm around climate change and other environmental disasters. If people feel threatened by the *news* of danger, then redoubling our efforts to spread the news will actually be counterproductive. As 'therapists to the world', we need to find ways of helping people become aware that they are in overwhelm, under the bedclothes with their fingers in their ears. We also need a parallel strategy of helping people reconnect with their innate love and awe for the other-than-human and more-than-human, so that they start to feel revulsion against their mistreatment.

These strategies are supportive rather than aggressive; in tune with the human capacity for Wild Mind, rather than driving it underground. However, they are hard to apply in a situation that screams *emergency* as soon as we let it into our consciousness. And, of course, change is not only a matter of individual or even collective consciousness. There are huge structures of power and money which necessarily oppose the wilding of the world, because any such process will destroy them. We may despise political leaders for their inaction around climate change, but few of them are fools: they know that capitalism can only survive through constant expansion, and that the consequences if capitalism abruptly fails would be disastrous – for humans at least – on a level similar to the consequences of climate change.

Some of therapy's current response to climate change is profoundly limited, and even actively unhelpful, because it restricts itself to the impact on human beings. This is not surprising, in that the same is true of the wider society, and of politicians in particular. But efforts to protect human beings alone from the consequences of our actions are both utterly ineffective on a practical level – human people cannot survive and flourish without other-than-human people – and a repetition of the exact behaviour that has caused this many-headed crisis in the first place.

Therapists need to listen to the voice of activists: 'Climate justice is social justice' – the efforts of the global North to make the global South pay for our destructive acts is one of the factors blocking an adequate response to the danger we are all sharing. Justice, in fact, is indivisible – real change for anyone means real change for everyone, since we are all intimately interconnected (Totton, 2023). If ecosystemic thinking means anything, this is what it means.

I believe that Wild Therapy has a role to play in the doubtful and difficult work of creating a new culture that can live well on the earth without damaging ourselves and other beings. There are many enormous obstacles to this work, and it isn't easy to see how it can succeed; it will involve both remaking our economic system and abandoning structures of domination and hierarchy that have been in place for millennia.

Fortunately, though, *things will happen of their own accord*, as newly emergent features of the complex web of being, not following any intention or plan. If a new culture is going to come into existence, then it must be already brewing, already cooking in many thousands of places around the planet; slowly assembling itself out of millions of local nano-acts of creativity and resistance. This may not be enough; but we can relax in the knowledge that it will be as good as it is possible for it to be. Accepting this will comfort us, stop us from wasting our time trying to control the future and at the same time show us our path, which is to envision and live the future we desire, and communicate this through our work with our clients.

## References

Feng, G. F. and English J. (1972) *Tao te Ching*. Aldershot: Wildwood House.

Le Guin, U. K. (1988) *Always Coming Home*. London: Grafton Books.

Macy, J. (1991) *World as Lover, World as Self*. Berkeley, CA: Parallax Press.

Macy, J. (1995) *Mutual Causality in Buddhism and General Systems Theory: The Dharma of Natural Systems*. Columbia: South Asia Books.

Samuels, A. (1993) *The Political Psyche*. London: Routledge.

Totton, N. (2021) *Wild Therapy: Rewilding Inner and Outer Worlds*. 2nd edition. Monmouth: PCCS Books.

Totton, N. (2023) *Different Bodies: Deconstructing Normality*. Monmouth: PCCS Books.

Voice 10

# Saving our children by bringing back beavers

*Eva Bishop*

We are at such a pivotal moment in human existence that you cannot separate the holistic context of climate breakdown and the loss of human habitat from the psychotherapeutic treatment of anyone remotely aware of the climate emergency. What it means to be human and our relationship with other life is changing.

Two decades of climate-driven career choices have led me to work on the return of the Eurasian beaver to Britain, a rodent once common across the northern hemisphere which is now helping to heal our desecrated landscape and restore the biodiversity to which humans have so recklessly laid waste. I do this because it feels like direct climate action. It adds meaning to desk-bound days and feels like I am contributing to one strand of climate resilience for future generations.

Beavers are a keystone species, the 'ecosystem engineer', modifying their habitats by building dams, digging canals, felling trees and opening up the riverside canopy to create a mosaic of beautiful, wild wetland habitats. In doing so they reverse decades of damage and drainage, and create homes for myriad species which return in great numbers to thrive in the complex landscape. The beaver's natural behaviour helps wildlife and people build resilience to drought, flood and fire, improves water quality, sequesters carbon and reinstates wild places – of vital importance to mind, body and soul.

But for beavers to thrive, we humans need to give them space, and there is unbelievable battle for ground everywhere you look within the conservation and land sectors. The lack of urgency or holistic context on occasion makes me want to scream. I hold this tension between rewilding for conservation's sake and farming for food production; but it's not one or the other, it's both! We all face a water crisis, we all face food shortage, we all face extinction.

Reintroducing beavers isn't just about saving one animal, nor solely about tackling biodiversity in crisis, it is about saving ourselves, envisioning a regenerative future and relearning our place as part of nature, giving her space to function and reconnecting with her in a way that we in the developed world or global north haven't known for hundreds of years.

DOI: 10.4324/9781003436096-30

Having a climate-aware therapist has been crucial to my journey as I've sought to make sense of a complex web of trauma and challenge. Climate has been a mast against which to pin the hardest decisions of transition while facing divorce, relocation, redundancy, parenting, climate anxiety and seemingly relentless grief. The climate emergency drives a constant pressure to be doing more. It is all I can see. So I have needed a therapist who understands the extremity of what we face.

I see my children's future, the need for soil, water and landscape recovery. This is why saving beavers is doing my bit for the climate crisis. They are a small creature, with a planetary message: if we step back, nature can help us solve multiple problems. But we need to shake the hubris – the human ego fighting for control – and adjust. We are not separate from wildlife and nature; how we treat wildlife is how we treat ourselves. And self-respect is intrinsic to the psychological journey. At planetary scale, respect and care must be our response to climate. Psychologists can help lead a cultural rewilding of a billion minds.

Beavers are only one story, not even *my* story, but they connect me with healing Earth. Climate-aware therapy has helped me wade through the depth and complexity of escaping convention and a marriage, supporting struggling children and understanding deep climate emotions. I've felt stuck, lost, despair, rage and unbelievable loneliness. But living with this and moving through it into a fully regenerative, connected place has started for me by connecting my home life, family relationships, work and purpose with climate, and I can see how much this sense of collective purpose could help so many others. I am in the fight of my life on my therapeutic journey, but it is deeply connected to the fight of all our lives in wider reality. For me, beavers offer some hope, for you it will be something else. But we must hold climate at the core of all we do. The more I love life, the more I care about our treatment of soil, of wildlife, and of my children's future. I want to thrive, and for that I need the planet to thrive.

Eva Bishop, UK

Chapter 16

# Transforming Our Inner and Outer Landscapes

*Leslie Davenport*

The dominant approaches to the frightening climate change communicated through the media and public education are insufficient. Our climate impacts have continued unabated, escalating to the level of existential peril. Depth psychology offers an important understanding to this (Mathers, 2020; Weintrobe, 2013) through bringing awareness to and transforming unconscious biases that have become the invisible drivers of climate change. We cannot address or understand the climate crisis without addressing the causal roots, and, to help with this, psychotherapists can remember the provenance of our work with the psyche. The Latin root of 'psyche' is *psukhē* and is translated as 'breath, life, soul'. It reminds us to open to the dimensions of the subtle, unconscious and transpersonal layers of experience. It is not enough to rearrange our thoughts, habits, and behaviours.

Our perceptions shape our beliefs, which become behaviours, creating lifestyles and norms, and impacting the environment – this circles back to impact us physically, mentally, and emotionally.

If we learn to reactivate dormant capacities that include creativity, contemplative wisdom, and somatic, collective, and ancestral ways of knowing, we can 'rewild' the mind (Totton, 2021) and address the root causes of our misaligned behaviours. Interior rewilding is an inside-out approach that we can apply while we also work outside-in by taking effective climate actions (Rust and Totton, 2019). These foundational shifts, when combined with polyvagal co-regulation (Porges, 2018), will translate into community action as they increase our ability to recognise our true role in the living biosphere. As we integrate these psychological tools into our interdisciplinary efforts to address climate change, we will significantly increase the probability of collectively creating a safer and healthier outer world.

Climate impacts that evoke eco-anxiety and grief – raging wildfires, rising seas, and more violent and destructive storms – communicate that something is terribly wrong, that life on our planet is dangerously out of balance. While efforts like reducing carbon emissions, halting the clear-cutting of forests, and implementing restorative agricultural practices are essential, we must also

DOI: 10.4324/9781003436096-31

restore the health of our inner landscape – our minds and emotions – and correct the distorted perceptions that have created our damaging lifestyles.

When we expect unbridled growth using finite resources, treat living systems as if they are commercial commodities, and systematically oppress whole segments of the population, it is only a matter of time before these destructive behaviours lead to chaos and collapse. Life is unable to thrive in diseased conditions. This view, however, is clouded by the complexities of the human mind that originated in our evolutionary psychology and have been perpetuated in our cultural worldviews and structures throughout history (Zhao and Luo, 2021). Psychotherapists can play a vital role by helping us un-learn our distorted beliefs and transform or 'rewild' our internal landscape (Totton, 2021) so that we make eco-wise choices that are in harmony with the living biosphere.

## Inner Transformation

Interior transformation calls on intuitive, creative, somatic, collective, ancestral, environmental, archetypal and emotional intelligence. In Western culture, these ways of knowing have largely become dormant. When we reactivate them, the result may be a route towards climate justice and restorative action. Without these more holistic ways of perceiving the world and each other, we find ourselves in a fragmented experience, a loss of empathetic relationality (Davenport, 2016) and we lose our ability to tolerate the painful unfolding of the climate crisis.

## Interior Rewilding Practice

Any practice that temporarily quiets our planning/analytical mind to create interior space for other ways of knowing contributes to our transformation. Practices like mindfulness (Kabat-Zinn, 2013), guided imagery (Davenport, 2009), expressive arts (Malchiodi, 2002), and focusing (Gendlin, 1982), support other means of perception. An essential component is being *actively receptive*: attending to whatever emerges from our interior ecosystem. This can be experienced as inner listening, inner seeing, a felt sense, the appearance of a dream figure or an impression. This is not about imagining something; rather, it is allowing what emerges to be perceived so that we can more clearly interpret the language of our inner world.

The integration of experiences through verbal expression is often helpful. Our deepest ways of knowing can be integrated with our more analytical mind as we grow more fully into what it means to be human. With practice, the ability to access subtle, creative and intuitive inner terrain and capabilities becomes as familiar and easy as our currently ingrained orientations. Taking time for regular practice is like forging a path that, once established, becomes easy and familiar territory to travel, and our brain's neural pathways take a different shape.

## Guerrilla Therapy

'Guerrilla gardening' (Davies, 2022) is where community members grow food and flowers on abandoned sites like vacant lots and roadsides. These efforts heighten community bonding and revitalise neighbourhoods. What starts out as a clandestine action often evolves into a community garden, with some supported by local governments. We can bring this spirit of guerrilla action to transforming our mental health by taking therapeutic support into areas of the community that are neglected and under-served and planting heirloom seeds of resiliency, beauty and nourishment in unexpected places.

There is another way that inner change spreads throughout our outer social circles. The process is as mysterious, hidden and essential as the mycorrhizal network that links individual trees through an underground web and quietly transfers nutrients to sustain the life of the forest (Simard, 2021). Polyvagal Theory describes an invisible network that enables us to go beyond self-regulation to co-regulation, nourishing the other human lives in our immediate biological and neurological network. Porges (2022) argues that our nervous systems synchronise with one another and allow us to nonverbally feel, empathise and share our states of being with each other. We are bioenergetically 'contagious' and when we cultivate interior rewilding, therapeutic self-regulation of our climate distress passes between people in this co-regulation process. It is a powerful alchemy transforming self-care into community care.

## Evolutionary Psychology's Hidden Drivers of Climate Change

We need to do more than re-engage our dormant inner capacities, we also need to address the internalised biases that shape our perceptions and actions. Once we recognise how these unconscious dynamics became dominant, we can consciously work with them, and cultivate a process of transformation that will help us move toward making more eco-wise choices intuitively.

## Bystander Behaviour

We gauge the reactions of others and the events around us in order to maintain our belonging/survival. When members of our group fail to react, we may take this as a signal that our response is not needed (Latané and Darley, 1970).

We see this today in our tendency to believe that if others are failing to respond to the threat of climate change, it is not necessary for us to respond either. In modern societies, we are inclined to assign the role of responding to danger to our political leaders – even while our faith in government has been eroding (Hickman et al., 2021).

## The Sunk-Cost Fallacy

The more we invest time, energy, or resources in an endeavour, the more likely we are to persevere, even if the outcome has damaging consequences. Historically, conserving resources may have been essential to our survival and abandoning an effort too soon could mean wasted energy. Staying the course also gave us an edge in competitions with other groups over resources (Arkes and Ayton, 1999).

This is often amplified by the loss-aversion bias: the fact that we experience the pain of loss twice as intensely as the equivalent pleasure of gain (Kahneman, 2013). This bias urges us to over-invest in ongoing commitments – such as our continued reliance on fossil fuels as a primary energy source, even in the face of decades of evidence of the destruction it is causing and the risk to all life forms, including ourselves. Bringing these unconscious dynamics into the light of awareness is a vital first step in personal transformation, and points to the rapidly growing importance of the mental health field as a valuable partner in forging climate solutions.

There are also capacities that are products of our early development that we can draw on to drive positive change. One of them is our tendency to engage locally in small work groups. When we are surrounded by people who are taking positive action to address the climate crisis, we are inclined to do the same (Legros and Cislaghi, 2020). This kind of social 'norming' is especially strong when we act collectively in response to a positive incentive: for example, improving the water standard in our town.

## The Architecture of Cultural Shifts

The collective depth transformation that climate disruption asks of us may seem daunting, or perhaps naïve and unrealistic in the face of the scale and complexity of the crisis. But in 2021, the University of Pennsylvania published a compelling research article in *Nature Communications* about the structural dynamics of widespread cultural shifts (Guilbeault and Centola, 2021). The study provided striking evidence that only 25% of the individuals in a social group need to make a shift before significant social change follows. Centola (2021) spent ten years developing a method for studying change in large populations, and he indicates that the social tipping points apply to any type of movement or initiative.

In the Global North we have improved public health and safety through shifts that are now integrated as cultural norms: eliminating indoor smoking, an international ban on chlorofluorocarbons, the universal use of seatbelts and women's right to vote. While these examples are less complex than the transformation that the climate crisis demands, they demonstrate encouraging successes in changing cultural norms that were previously viewed as 'just the way things are'.

A central contribution of climate psychology work is helping clients build their emotional resilience which has frequently been measured by our ability to bounce back to baseline from a challenging life disruption (Graham, 2018). This definition needs to be transformed in the face of the climate crisis and climate aware therapy practice, because there is no longer any going back to a predictable baseline, for us or the planet. Climate chaos is on an escalating trajectory, so our current challenge is to develop a flexible emotional agility that allows us to remain present, open-minded and empathetic in the face of the growing distress in our warming world.

It is best summed up in the words of Rainer Maria Rilke, 'May what I do flow from me like a river, no forcing and no holding back …' (in Barrows and Macy, 2005: 65).

## References


Arkes, H. R. and Ayton, P. (1999) The Sunk Cost and Concorde Effects: Are Humans Less Rational Than Lower Animals? *Psychological Bulletin*, 125(5): 591–600. 10.1037/0033-2909.125.5.591.

Barrows, A. and Macy, J. (2005) *Rilke's Book of Hours: Love Poems to God*. New York: Riverhead.

Centola, D. (2021) *Change: How to Make Big Things Happen*. Boston: Little, Brown & Co.

Davenport, L. (2009) *Healing and Transformation through Self-Guided Imagery*. Berkeley: Celestial Arts.

Davenport, L. (2016) Mindful Advocacy: Imagery for Engaged Wisdom. In: L. Davenport (ed.) *Transformative Imagery: Cultivating the Imagination for Healing, Change and Growth*. London: Jessica Kingsley Publishers, pp. 236–247.

Davies, R. (2022) What Is Guerrilla Gardening and How Does It Help the Climate, *Euronews*. Available at: www.euronews.com/green/2022/08/25/what-is-guerrilla-gardening-and-how-does-it-help-the-climate [Accessed 26 May 2023].

Gendlin, E. (1982) *Focusing*. New York: Bantam Books.

Graham, L. (2018) *Resilience: Powerful Practice for Bouncing Back from Disappointment, Difficulty, and Even Disaster*. Novato: New World Library.

Guilbeault, D. and Centola, D. (2021) Topological Measures for Identifying and Predicting the Spread of Complex Contagions. *Nature Communications*, 12(4430): 1–9.

Hickman, C., Marks, E., Pihkala, P., et al. (2021) Climate Anxiety in Children and Young People and Their Beliefs About Government Responses to Climate Change: A Global Survey. *Lancet Planet Health* 5(12): e863–e873.

Kabat-Zinn, J. (2013) *Full Catastrophe Living*. Boston: Little, Brown Book Group.

Kahneman, D. (2013) *Thinking Fast and Slow*. New York: Farrar, Straus and Giroux.

Latané, B. and Darley, J. (1970) *The Unresponsive Bystander: Why Doesn't He Help?* Hoboken, NJ: Prentice Hall.

Legros, S. and Cislaghi, B. (2020) Mapping the social norms literature: An overview of reviews. *Perspectives on Psychological Science*, 15(1): 62–80.

Malchiodi, C. (2002) *The Soul's Palette: Drawing on Art's Transformative Powers*. Boulder: Shambhala.

Mathers, D. (ed.) (2020) *Depth Psychology and Climate Change: The Green Book.* Oxon: Routledge.

Porges, S. (2022) Polyvagal Theory: A Science of Safety. *Frontiers in Integrative Neuroscience*, 16, 871227: 1–15.

Porges, S. (2018) *The Polyvagal Theory in Therapy: Engaging the Rhythm of Regulation.* New York: W. W. Norton & Company.

Rust, M. J. and Totton, N. (eds) (2019) *Vital Signs: Psychological Responses to Ecological Crisis*. Oxon: Routledge.

Simard, S. (2021) *Finding the Mother Tree: Discovering the Wisdom of the Forest*. New York: Knopf.

Totton, N. (2021) *Rewilding Our Inner and Outer Worlds*. Monmouth: PCCS Books.

Weintrobe, S. (2013) *Engaging with Climate Change: Psychoanalytic and Interdisciplinary Perspectives.* Oxon: Routledge.

Zhao, J. and Luo, Y. (2021) A Framework to Address Cognitive Biases of Climate Change. *Neuron*, 109: 3551.

Chapter 17

# The Spiral of The Work that Reconnects

*Chris Johnstone and Rosie Jones*

## Introduction

How often have you heard the phrase 'we can't change the world'? If that were true of our collective impact, we wouldn't have a human-made climate crisis. But the psychological reality that it feels true to many people is both an obstacle to climate action and an amplifying factor for distress about global issues. A survey commissioned by the Mental Health Foundation found powerlessness to be the most common emotional response to world problems, with many people also feeling angry, anxious and depressed (BBC News, 2007). Could we develop an approach to therapeutic practice that helped people grow and apply their sense of agency in response to concerns about the world? And could this type of practice strengthen our collective transformational resilience in a time of worsening world conditions?

This chapter describes such an approach. Known as *The Work that Reconnects* (TWTR), it has engaged participants in at least six of the seven continents and touched the lives of hundreds of thousands of people. Initially developed as an approach to group work, its application has spread to find its place in activist gatherings, conversations of mutual support, therapy consultations, university courses, meetings of faith-based groups, works of art and online programmes. A recent outcome survey showed benefits included increased motivation to act for positive change and strengthening of the belief that we can make a difference, with participants feeling less overwhelmed or defeated about their concerns for the world after engaging with this approach (Johnstone, 2023).

In this chapter we focus on a central feature of The Work that Reconnects – a sequence of four elements known as 'the spiral' – exploring how we can apply this in a range of settings to nourish our capacity to act for positive change and help others do this too.

## Beginnings

In the late 1970s, US academic and activist Joanna Macy developed an approach to group work known as 'despair and empowerment work'. With

DOI: 10.4324/9781003436096-32

*Figure 17.1* The Spiral of The Work that Reconnects.
(Macy and Johnstone, 2022).

parallels to the way grief work offers supportive space for people to engage with the grief of loss, Macy's despair and empowerment work invited people to acknowledge their inner reactions to disturbing events in the world. Through partnered and group sharing processes, participants explored and expressed concerns often difficult to give voice to in more everyday conversations. A series of experiential practices and teaching inputs then supported participants in building change-making capacity and developing their role in acting for positive change. As so many experienced this work as energising and empowering, it quickly spread from its origins in the US to many countries around the world. In their book *Coming Back to Life*, first published in 1998, Joanna Macy and her colleague Molly Young Brown introduced two important developments of this work (Macy and Brown, 2014). The first was a new name that was catching on to describe this approach – *The Work that Reconnects.* The second was the spiral framework for offering the work that moves through the four successive phases of Coming from Gratitude, Honouring Our Pain for the World, Seeing with New Eyes (more recently referred to as Seeing with New and Ancient Eyes) and Going Forth (see Figure 17.1).

## Four Settings for the Spiral

The guiding structure of these four elements is referred to as 'the spiral' because each repetition might reveal something new, whether a deepening of the familiar or an unexpected new angle. As we review, each stage of the spiral can be linked to evidence-based psychological interventions. Yet while each element brings benefits by itself, the power of the spiral is rooted in the way they act together as a whole that is more than the sum of its parts. It is the sequence of elements, and the synergy between them, that we focus on, because once this form becomes familiar, it can be used in many situations. We will look at applying the spiral in the following four contexts.

*Personal Practice*
You have just watched a disturbing documentary or read an article you found upsetting. You are wondering how you can support and steady yourself, as you take the information in and consider your response.

*Therapeutic Encounters*
You are working with a client in therapy who is deeply concerned by world events and not sure how to cope with the distress they feel. You might also feel triggered by this topic yourself and so consider responses both to support your client and to help you maintain your therapeutic presence.

*Peer-based Support*
A friend is deeply troubled by a concern you have too. You ask whether they might like to explore the spiral form together as a framework for mutual support and shared learning. It could be something you do together, or you might ask others to join you too.

*Transformative Groupwork*
You take part, as a participant or facilitator, in a group-based journey round the spiral at a Work That Reconnects workshop.

In each of these settings, the spiral offers a pathway to follow and options to explore. It becomes somewhere to turn if we, or those we are working with, are stirred by concerns about world events and want to tap into sources of strength or inspiration to respond. When someone feels overwhelmed, anguished or despairing, this spiral form, with its associated insights and practices, supports a rootedness that is steadying and a sense of possibility that helps turning points become more likely.

The book *Active Hope*, first published in 2012, introduced a way of working with the spiral as a personal practice, the book guiding the reader on a similar journey to that found in a workshop (Macy and Johnstone, 2022). An unanticipated development was the spread of 'Active Hope book groups', where people met to discuss the book and share practices it describes. Launched in 2021, a free video-based online course at www.ActiveHope.Training offers another pathway of engaging with the spiral, either as a solo journey or as a shared adventure in peer-based groups. 'Circles of Active Hope', where groups meet regularly to follow the spiral journey, either guided by facilitators, or in self-directed groups supported by the online course or book, have emerged as new forms for The Work That Reconnects to happen.

While many benefit from journeying round the spiral by themselves, there are significant added gains from doing this work with others. People we interviewed told us they feared being labelled as depressing company if they shared the depths of their concerns about the world in public. Yet when alarm about global issues is hidden from view, or perhaps raised only in individual therapy, it is possible for everyone in a group to feel distressed, but each to believe they are the only one. A common feedback comment after workshops is 'I never knew so many others felt like me'. A function of this work is to provide supportive contexts and a sense of solidarity as we bring our concerns into the open, so that we can then look at what helps us respond to them. Facilitated workshops can take people on a deeper dive, with a greater level of support, a wider range of activities and more intensive forms of practice. The transformative alchemy of community that emerges from this is a much-needed healing factor in our times. When the problems of our world feel too much to face by ourselves, the expression 'I can't, we can' conveys one of the essential meanings behind the phrase 'The Work that Reconnects'.

In any form of therapy, we can draw upon the spiral or point people towards it. Just asking the question 'what might it look like if I applied the spiral here?' can help us access this guiding framework to follow. While applying The Work that Reconnects in a therapy session, personal practice or self-guided group differs in important ways from the experience of an intensive workshop, each of these settings can invite steps on a similar journey. These modalities can mutually reinforce each other and offer complementary forms of transformational support.

## Gratitude

Gratitude practices offer a gentle, attractive way to begin that tends to put people at ease, generating psychological warmth, trust and a sense of safety. Starting this way helps participants feel better resourced to face the more challenging material in the second part of the spiral.

Research shows gratitude work brings many mental health benefits, not only improving mood but also strengthening resilience (Wood, Froh and Geraghty, 2010). In a review exploring the potential impact of gratitude interventions in psychotherapy, Emmons and Stern (2013: 854) conclude: 'Gratitude is a key, underappreciated quality in the clinical practice of psychology, its relevance deriving from its strong, unique, and causal relationship with well-being, as well as its dynamic healing influence …' Gratitude practices have also been shown to strengthen pro-social behaviour, with significant effects confirmed in a recent meta-analysis of 91 studies (Ma, Tunney and Ferguson, 2017).

One of the tools often used in workshops is the sentence starter, which points participant's flow of words in a particular direction. As gratitude involves both appreciation of something valued, and recognition that someone or something has played a role in this, simple sentence starters such as 'I'm grateful for …' and 'I'm grateful to …' draw out these different sides.

Whether in paired listening exercises, or as group rounds, listeners just give their attention, opening a free space for each speaker to explore where each starting point might lead. A thanksgiving circle, guided by the words 'I'd like to thank ...', can be a deeply moving process, reminding all present of the network of support sustaining them.

An alternative to 'I'm grateful for ...' is 'I love ...', which invites the speaker to identify where or when their heart connects in a loving way, whether with people, places, activities, values, foods, music, qualities or whatever else. This is an uplifting activity that also orientates people to what they care about and wish to protect; it leads quite naturally into the second stage of the spiral.

Grateful appreciation can also be practised through guided meditations, which might reflect, for example, on the evolutionary marvels of the human body and the senses, or on nature. Deep, appreciative noticing can be a novel and profoundly nourishing experience.

**Try this:** Either as a personal reflection or as a partnered listening exercise, see what words follow these starting points. Whenever you're not sure what to say, say the sentence starter again and see what comes.

1 I love ...
2 I'd like to thank ...
3 When I feel grateful, I'm more likely to ...

## Honouring Our Pain for the World

If you were to treat a visitor as an honoured guest, how would you respond if they knocked on your door? Would you welcome them in? To honour someone is to treat them with special attention and respect, recognising their value. A starting point for honouring our pain for the world is just to be aware of our response when we feel it knocking at the door of our attention.

Pain for the world is an umbrella term for the range of ways people experience distress about disturbing world events. Emotions commonly felt here include alarm, anxiety, grief, sadness, guilt, anger, dread, overwhelm and shame. Whether we open our door to such feelings, or try and shut them out, will depend in part on our beliefs about their value. If we view them as life-preserving signals alerting us to threats, it makes sense to give them our attention. Distress on behalf of someone or something else is also a signal of relatedness; it shows not just that we've noticed, but also that we care. In the second stage of the spiral, we honour these feelings by making room for them, valuing our heartfelt responses to crises in the world as expressions of our awareness, connection, and concern.

**Try this:** As before, either as a personal reflection or in a partnered listening exercise, see what follows these sentence starters:

1 When I look at the future we're heading into, concerns I have include ...
2 Some feelings that come when I think about this are ...
3 What I do with these feelings is ...

In Motivational Interviewing, the process of 'evoking change talk' is a central mechanism for helping people change. Change talk is where someone speaks in a way that supports a shift in a particular direction, such as expressing a desire or need for things to be different, making a commitment to this and identifying steps they might take. Research links expression of change talk to increased likelihood of following through with action (Amrhein et al., 2003). Several of The Work that Reconnects practices, including the sentence starters above, invite expression of change talk. When someone hears themselves describing strong feelings about an issue, that can be an important step in talking themselves into addressing it.

When people open the door to their pain for the world, grief is often among the feelings that come up and sometimes it can feel overwhelming. In her study of what helps people deal with loss, Hone (2017: 213) writes: 'Public Mourning rituals, such as funerals, have a clear purpose. By gathering people together around the bereaved, they help mourners strengthen their bonds and re-enter the social world after a major loss.' In this time of climate crisis, where are the public mourning rituals for what is being lost in our world? Research shows that personal rituals can also play an important role in the grieving process (Norton and Gino, 2012). The Work that Reconnects offers practices to meet this need, one of them being the *Cairn of Mourning*. We invite you to try a version here adapted for personal use, or to consider inviting clients to try this, or something like this, when grief linked to world issues comes into view.

**Try this:** A Personal Cairn of Mourning
What is being lost in our world that you mourn for? While out for a walk, reflect on this question and find a small object you can bring home that symbolises this. Each time you do this, place your object in a special place you've set up in your home or garden to mark significant losses you mourn. Over time, such objects add together to create a cairn of mourning to honour the grief you feel.

## Seeing with New and Ancient Eyes

The first two stages of the spiral are about taking in the reality of what is happening in our world, in both its beauty and its horror. The next two parts address the challenges of finding our part to play in response and drawing in support to help us do this. With honouring our pain for the world, the starkness of our situation is faced; it might at times seem too big or too far gone for us to make any difference. A task of the third stage of the spiral is to seek out perspectives and resources that support a shift in our sense of possibility, encouraging and inspiring our engagement with the purpose of acting for our world.

There are two aspects to this third stage of the spiral. The first is about inviting a willingness to step beyond the familiar and look in different ways, experimenting with fresh perspectives fed by our imagination, creativity, a range of sensory inputs and sources of wisdom both ancient and new. The second aspect involves a transformation that lies at the heart of The Work that Reconnects: from seeing ourselves as separate individuals to more deeply embracing the experience and understanding that we are part of the living Earth. This pivotal recasting of our view of reality doesn't just happen at a cognitive level – when well-developed it also involves a shift in consciousness.

When someone is a member of a team they love being part of, they think in terms of 'us', with a collective identity linked to feelings of loyalty, purpose and belonging. The consciousness of experiencing ourselves as part of the living Earth is similar – it is as though we are part of the larger team of life. When this orientation lands within us, it can be a powerful guiding impulse; it is something we might know in our bones and feel in our heart. While the different elements of the spiral act together to strengthen this deep knowing of our 'part-of-ness', the third stage brings it more clearly into view. This recognition that we are part of the Earth and deeply connected with life is an ancient one found in many indigenous wisdom traditions; acknowledging that, this stage of the spiral is increasingly referred to as 'Seeing with new and ancient eyes'.

Gaia Theory, which is based on living systems thinking, is an important source for this part of the spiral, particularly in relation to the way we think about pain for the world. With the approach of systemic family therapy, a family is viewed as a living system that acts through its members. If a close relative is injured, a family may also feel through its members, emotions of concern serving a function in mobilising a caring response. Our feelings for our world can be viewed in a similar way. Anxiety or grief about global issues can be thought of as Gaian emotions, as our world feeling through us. When these feelings stir us to address these issues, perhaps we can think of this as our world acting through us too.

How long does a family live for? That question invites a widening of timescape beyond that of our own lifetime. In some cultures, it is not unusual to have meaningful relationships with ancestors or those not yet born. A fertile area of activity in The Work That Reconnects is 'Deep Time work', where we cultivate our felt sense of connectedness with those who live in different times. By growing our identification with a community of life that extends both back and forward in time, it is possible to experience solidarity with, and support from, both ancestors and future beings. Research identifies social support as a key protective factor in times of stress and even imagined support has been shown to bring benefits (Jakubiak and Feeney, 2016). With that in mind, we invite you to try this practice for looking in different ways that can strengthen access to a wider network of support.

**Try this:** Widening Circles
This practice allows people to express their views and feelings about a topic while also experiencing different perspectives, and larger contexts, with respect and empathy. In a workshop or peer support context, people sit in groups of three or four. Each chooses a particular topic that concerns them and then speaks about this from each of the following four perspectives, while the rest of the group listens.

1 From their own point of view, including their feelings about it.
2 From the perspective of someone holding opposing views on the issue, speaking from 'I'.
3 From the viewpoint of an 'other than human' being that is affected by this issue.
4 From the voice of a future human who will be affected by the choices made now on this issue

This process can be adapted for personal use as a self-reflection exercise, imagining yourself moving between the different roles and writing a page from each perspective.

## Going Forth

The fourth stage of the spiral focuses on helping people find their part (or parts) to play in responding to concerns they have. Called 'Going Forth', this stage offers tools, insights and contexts of support that strengthen people's ability to listen for what calls to them, both in terms of the issues they want to address and the responses they want to make. When we are acting from our interests, passions and enthusiasms, and drawing on strengths, skills and experiences that are uniquely ours, we are more likely to bring our best to whatever we do.

We can approach the process of taking action by looking at the story we would like to live in. One way of looking at our current climate crisis is that it is just one element in a larger story of planetary decline that also includes extreme inequality, war, mass extinction of species and habitat destruction. We call this story *The Great Unravelling*. A powerful driver of unravelling is the dominance of another narrative we call *Business as Usual.* The Work that Reconnects is rooted in a third story we call *The Great Turning*, where crises can become a turning point marked by an inspiring collective transition to a life-sustaining society. The Going Forth part of the spiral invites people to consider how we can each find our part to play in this third story.

Practices at this stage may involve clarifying intentions, working through doubts and obstacles, identifying resources, both inner and outer, and hearing support and encouragement from others, including ancestors and future generations.

The movement around the spiral of facing reality, choosing the story (and values) we want our lives to express and then identifying specific steps we can take, aligns closely with the approach of Acceptance and Commitment Therapy (ACT). In their book describing this approach, Hayes and Lillis (2012: 245) write: 'Instead of waiting for a solution, ACT focuses on living a life passionately connected to that individual's values, now.' We can apply that approach to the climate crisis and other aspects of The Great Unravelling. In looking at how we connect with a larger story we have our heart in, perhaps the most important part of The Great Turning is turning up with an intention to play our part. These sentence starters are useful prompts here.

**Try this:** As before, either as a personal reflection or in a partnered listening exercise, see what follows these sentence starters:

1 When I look at the future we're heading into, what I deeply hope for is …
2 A part I'd like to play in support of this is …
3 A step I'll take towards this in the next week is …

## Activating Hope

Active Hope is something we do rather than have. It involves identifying what we hope for and then taking steps towards these hopes or steps that help them become more likely. When we activate hope, it happens through us. Through our choices and actions, we play our part in a larger story that does change our world. The spiral is a tool and transformative pathway that not only helps us do this, but it strengthens our ability to help others do this too.

## References

Amrhein, P. C., Miller, W. R., Yahne, C. E. et al. (2003) Client Commitment Language during Motivational Interviewing Predicts Drug Use Outcomes. *Journal of Consulting and Clinical Psychology*, 71: 862–878.

*BBC NEWS* (2007) | *UK* | *World Troubles Affect Parenthood.* Available at: http://news.bbc.co.uk/1/hi/uk/7033102.stm

Emmons, R. A. and Stern, R. (2013) Gratitude as a Psychotherapeutic Intervention. *Journal of Clinical Psychology*, 69(8), August 2013: 793–867.

Hayes, S. C. and Lillis, J. (2012) *Acceptance and Commitment Therapy*. Washington DC, American Psychological Association. Kindle edition.

Hone, L. (2017), *Resilient Grieving*. New York, The Experiment Publishing, p. 213.

Jakubiak, B. K. and Feeney, B. C. (2016) Keep in Touch: The Effects of Imagined Touch Support on Stress and Exploration. *Journal of Experimental Social Psychology, 65*: 59–67.

Johnstone, C. (2023) Building Change-making Capacity: Active Hope Training. In: L. Aspey, C. Jackson and D. Parker (eds) *Holding the Hope*. Monmouth: PCCS Books, pp. 101–110.

Ma, K. L., Tunney, R. J. and Ferguson, E. (2017) Does Gratitude Enhance Prosociality? A Meta-analytic Review. *Psychological Bulletin*, 143(6): 601–635.

Macy, J. and Brown, M. Y. (2014) *Coming Back to Life: The Updated Guide to The Work That Reconnects*, 2nd edition, Gabriola Island, British Columbia: New Society Publishers.

Macy, J. and Johnstone, C. (2022) *Active Hope: How to Face the Mess We're in with Unexpected Resilience and Creative Power*, 2nd edition, Novato, CA: New World Library.

Norton, M. I. and Gino, F. (2014) Rituals Alleviate Grieving for Loved Ones. *Journey of Experimental Psychology: General*, 143(1): 266–272.

Wood, A. M., Froh, J. J. and Geraghty, A. W. (2010) Gratitude and Well-being: A Review and Theoretical Integration. *Clinical Psychological Review* 30(7): 890–905.

Voice 11

# Wings of hope

*Will Baxter*

When I discovered wildlife photography, I was astounded to find two things I love the most – wildlife and photography – in one. Behind my house there is a field with a river going through it. There is a variety of animals from foxes to deer to woodpeckers and kingfishers. And in the long grass there are mice scavenging, making it a prime hunting spot for barn owls. On the first day I saw the barn owls there were three of them. A barn owl looks like a wastepaper bag floating on the breeze, flying so silently. I did not get a shot on the first night, but I went back for 20 or so more days and took around 100 shots until I got this photo.

Barn owls are one of the most extraordinary and remarkable creatures. They are one of the only birds that can fly completely silently. Barn owls are just like us; they are highly sensitive and incredibly intelligent creatures. They also help out farmers by eating rodents and providing free pest control. But due to climate change there is an increase in summer storms and there are unpredictable changes in weather. And this is not the only thing that is posing a threat to barn owls. The hot and dry weathers are impacting the growth of grass meaning there are fewer grassland areas for barn owls to hunt. This is a big issue because there is not enough food, making some barn owls starve. Having a world without barn owls will not just affect the farmers but will affect the children of this planet. Barn owls have been roaming these lands for hundreds and hundreds of years and seeing these beautiful birds go will be horrific. And it will not just affect my generation but for generations to come. It would make me feel disappointed that we had the chance to save this majestic bird and we didn't take it. It is not the environment that is killing this bird, and killing hope, it is us.

Will Baxter (Age: 14), UK.

DOI: 10.4324/9781003436096-33

*Plate 11.1* Will's photograph depicts both the beauty of the bird as well as the love of the child for nature. It is hard to hear the silence of the flight in this photograph, which is his intention.

# Section Five

# Community and Social Approaches

Chapter 18

# Beyond the Ego and Towards Complexity through Social Dreaming[1]

*Julian Manley, Wendy Hollway and Halina Pytlasinska*

## Introduction

What can social dreaming bring to our thinking about living in a world beset by the effects of climate change and associated anxieties, such as fear, guilt, shame and resignation? Social dreaming is not used therapeutically but may often serve a therapeutic purpose. As the truth of the effects of climate change dawn upon people, it becomes difficult to focus on significant aspects of the problem within the search for a 'solution'. Likewise, during a pandemic that obliged people to change their routines, embedded 'givens' became inadequate.

Such requirements to significantly change lead us to question the value of individual efforts to deal with the climate emergency. During the pandemic, when individuals were no longer flying, planes flew empty anyway. The individual's electric car is built on the back of systems of slavery and extraction. These conundrums are such that it is easy to see how someone might fall into a state of overwhelming anxiety and desperation. In these circumstances, the practice of social dreaming can help by introducing alternative forms of expression and different uses of language that can be shared with others in solidarity and compassion (Manley, 2020).

## What Is Social Dreaming?

In social dreaming, people gather to share their dreams to a 'matrix', a liminal space for the creation of new thinking where an individual's dream becomes a 'social dream' through association. Participants in a social dreaming matrix are invited to share a nighttime dream with others in the matrix. Others can follow by sharing another dream that may have a link or connection to a previous dream. Alternatively, they may offer an association triggered by previous dreams and associations. New knowledge is co-created through these images and their affects as they accumulate and form a collage of image-affects infused with emergent meaning that is allowed to 'simmer'. Interpretation of dreams is not part of the social dreaming. This enables

DOI: 10.4324/9781003436096-35

abductive enquiry, where possibilities are added rather than foreclosed (Long and Harney, 2013).

In this way, the method itself mirrors the need for thinking in complexity that is demanded of the world in facing up to climate change. It also honours participation and removes any sense of judgement. A tenet of social dreaming is that individuals are not the focus because social dreaming emphasises reflection on the dreams as entities in themselves, not belonging to the dreamers who offer them. There is deliberately no focus on social dreaming. The dreams and associations are allowed to 'float'. As befits 'floating', the image-affects of social dreaming can create different configurations of Winnicottian potential (Winnicott, 1991). These are sufficiently random as to avoid a focus that would lead to interpretation and, along with that, reductionism and foreclosure of developing thoughts and feelings. Simultaneously, they are sufficiently contained and non-random as to bind the individual dreamers into a collective commonality that provides cohesion while respecting difference. The knowledge that emerges makes sense to the shared space of the matrix. Such is the unfocused focus of social dreaming.

Social dreaming provides a means of expressing complexity to people who might otherwise need to break up climate ecology into objectified fragments of thought. Any fragmentation would fail due to the whole's inseparability which characterises its core and its essence. The vast ecological system that governs our lives and is threatened by climate change is a complex whole that resonates with the complexity of the social dreaming matrix. Such complexity has been described as a 'hyperobject' by Timothy Morton (Morton, 2013; Manley and Hollway, 2019). Climate change as hyperobject cannot be 'seen' in its entirety but is noticed through its various manifestations, such as a flood or a wildfire. These merely provide clues to the whole, not a complete vision, which is intuited through the creative imagination.

Social dreaming enables an imaginative vision of the whole through 'condensation' (Freud 1991 [1900]) of images and multiplicities in a single dream which is further multiplied in the collection of dreams and associations produced in the matrix. The multiplier effect is enhanced through combining a severality of dreams and associations, received by participants as a rhizomatic process shared in the matrix and post-matrix discussion.

## Not a Talking Therapy

The therapeutic effects of social dreaming can be described as the relief gained from the shared and contained expression of ideas, thoughts and feelings which are largely unavailable through habitual discourse. This is sometimes referred to as allowing the 'unthought known' (Bollas, 1987) to emerge in the dreams and associations of the matrix.

Social dreaming differs from traditional psychotherapy in how the dream matrix emphasises the social, cultural, relational and environmental aspects of the psyche. As Serra Undurraga (2023) has described, many forms of psychotherapy assume subjectivities of the individual as self, as if the individual were wrapped up in a body considered to be impermeable to the environment. Social dreaming, with its collage of image-affects and its emphasis on shared dream material, contrasts with the idea that the individual is in possession of a dream. In this way, by completely removing the individual self from the shared understanding of the associative unconscious in social dreaming, the matrix is far removed from traditional psychotherapy. The traditional assumption is true even in gestalt and transpersonal psychology, where, despite an emphasis on 'beyond ego' or holism or interconnectedness, the purpose of these therapies always supposes a return to the mental health of the individual.

Many participants express gratitude, calm, a newly found centredness, joy and wonder. They feel bonding and solidarity with others in the matrix, extending to identification with the non-human (brought closer to the human in heterotopian dream images where the impossible is made possible (Manley, 2019: 33[1])).

Despite these therapeutic possibilities, social dreaming could never be considered therapy. It occupies the borderline between therapy and non-therapy, a non-clinical place where healthy co-creation can happen.

## A Collage of Affects: Examples from Social Dreaming During the Pandemic

This section investigates some of these image-affects and collages as they emerged in social dreaming work held online during the pandemic.[2]

In these sessions, the interweaving of dreams and associations[3] created alternatives that tackled existential questions about social and personal identity, our place in nature and social systems. For example, an association to the dreams brought up the idea of going up a canal through a series of 'locks', making progress in *lock*down, so to speak, but also how this human-created technology must work in harmony with nature:

> *you have to go into this lock in the boat, like this kind of manmade area, and you wait, and you have to wait for the sea or the canal water to rush in ...*

The association was made in reference to several water dreams emerging in the matrix. Each dream or association has multiple potential meanings in conjunction with the matrix as a whole, which cannot be fully expressed in writing. The 'locks' association lacks the richness of the collage of image-affects of the here-and-now of the matrix. Ultimately, this very inexpressibility of the matrix is its *raison d'être*. As authors, we do our best to express the complexity of the social dreaming matrix, even when this complexity is the reason for using the language and feeling state of social dreaming in the first place! In this matrix, the associated water dreams spoke of abandon or freedom by swimming in water and different states of water, still and contained, as in the swimming pool, and wild and free, as in the open sea.[2]

A similar feeling of abandon or freedom was expressed in several flying dreams, and brought together in a single association describing being 'catapulted' from the back of a boat and the intense emotion of freedom experienced in the memory:

> *I went back on the boat and flew and was catapulted up and, talk about, you know, fear but then this elated but moreover, not so much the excitement, but when you reach high, high, high, high, high, it was so silent. And I remember I could hear some birds, that at one point, there was nothing. It was literally the one of the most intense silences I've ever felt, I've ever heard, I've ever sensed …*

What matters here is the description of the affect of the association. Affect is prioritised over fact (the technical term 'parasailing' does not matter here). In the social context of the pandemic, it is possible to share the emphasis and value given to being 'high, high, high, high, high', feeling the fear, and the 'intense silence' of the experience. For participants of the matrix at this moment, this is the important thing: here the 'answer' of the matrix to the question of the lockdown and the pandemic is an affect that affords a re-evaluation of the importance of freedom. This idea is held in the exhilaration of the sensation when combined with nature as symbolised by water and air. Once this is realised, it is available for thought and use: the unthought known has been thought.

The matrices also engaged participants in nuanced contemplations of self and identity; they questioned their identities and by implication their positioning in the world as human beings through the multiple images and associations brought about by dreams. For example, dreams of being located above and below tables demonstrated, through the interconnectivity of the matrix, clear reference to a below-the-surface unconscious, and an above-the-surface consciousness. Another connected thread of dreams and associations related to clothes and costumes. The world in which participants had formed their self-awareness and identity appears to be on the verge of collapse. Old ways and systems do not appear to be guaranteed anymore. New thinking is required, something beyond getting back to normal or 'building back better'. The following association to dreams of tables and alternative possibilities of perception illustrates this core challenge to identity:

> *There's a recurring theme of tables and then this talk about being covered up or seeing another reality brings to mind when I was a little boy I used to, I didn't hide, but I used to just go under the dining room table. I enjoyed being under the table and looking at the world from that perspective. My daughter now still has that table, and it has my name written underneath it.*

There is a suggestion here that 'under the table' the speaker's daughter can find the father's youthful identity and that this may be more authentic and more useful in the current situation than any 'above table' adult consciousness. The speaker's adult perspective and adult identity are questioned, a reminder of the children's school strikes in protest against adults' climate inaction, personified by Greta Thunberg. In this way, the matrix looks for reactions to the current crises through an existential searching of the self, as opposed to returns to past norms and the identities associated with those norms. The risk of reaching for new ideas ('under the table', in the unconscious, a semi-hidden place) is expressed in the following childhood association connected to the dining-room table dream by the matrix, where the potential warmth from beneath the table ignites into hurtful burns:

> *When I was nine, I was playing cards with my siblings and cousins and in Spain in the little, in the villages, we had a big tray where you would put embers to keep warm. There was no central heating when I was a child, so under the bed, under the table, you would put this tray with the embers and then we had the tablecloth over the table, so we were playing cards and suddenly I started feeling this heat coming up my leg and I realised that my pyjamas were on fire. So, some lower part of one of my legs was badly burnt and I had to be taken to hospital.'*

In this way, the matrix rejects a nostalgic return to childhood identity. There is no feeling of a flight from adult reality in the matrices, or a regression into childhood. The sense of questioning the core of our identities, without personalising them, is nevertheless present throughout. The image of the table was further developed in the matrices by the addition of the idea of a new choice of costumes, part of that table, as in the following dream:

> *I had a dream last night and it was in a friend's house, we were having dinner but under the dinner table there was, I think it's called rack, like the thing like the hangers that we have in stores and clothes are hanging there. So, the down part of table was that and you could touch if you like to those pieces of coats. And at some point during the dinner, a very close friend of mine, she was sitting next to me, she said she needs to go out. And for that she needs my shirt. So, I took off my shirt. I gave it to her. She tries it, tries it on. And I think then she leaves.*

The feeling that a new costume, as yet untested, could be understood as a new identity was emphasised throughout the matrix. Of the many possible meanings of these below-the-table costumes, we offer two. The linear logic of the dream as it is related here tells that the shirt is given from her back to a close friend, in order to go out beyond the safety of the indoors dinner table. An obvious association is to the saying 'I would give the shirt from off my back', alluding perhaps to the altruistic spirit of lockdown communities and the possibility of new levels of sharing, including in the matrix. However, the shirt was plucked from the table, from the rack, no longer a coat but something closer to the skin, to the self. This association was supported by the wider matrix data. For example, in the following dream different costumes are tried on and linked to the wearer through references to teeth, eyes and bones, in other words the essential durable parts of the body and the self, covered in a new identity, yet subject to mask wearing:

> *[they had] fantastic, unusual costumes on and the one I can just give an example is that someone had like a sheet over their head. And their teeth were sewn into the sheet. So, you could see their mouth because they had sewn the teeth into the sheet. And then they had drawn like a skeleton head or a head-like skeleton on the sheet. And maybe they had cut open the eye holes.*

This is directly linked to mask wearing during Covid-19 in the following association, so replete with potential associations that we can only offer it for reflection and pass on:

> *I was walking with two friends and there was this woman who had a mask on, but she had put on lipstick like a purse, pursed lips on the mask.*

Through these examples, there is a sense of looking into complexity and a refusal to be satisfied with simple ways out of the crisis. The way the matrices invent a new reality – a willingness to creatively imagine an alternative life that is new and vital, rather than a return to the old ways – can be a valuable addition to the widely-discussed need for a socio-economic transformation. Although many people might be aware of the need for an alternative life, perhaps they can hardly articulate it, in the sense of it being part of the 'unthought known'.

## The Ecological Vista of Social Dreaming

Traditional psychotherapy emphasises the self and the ego, usually closely entwined despite the use of dream analysis and the couch to remove eye contact. This is also true of therapeutic group work using dreams (Ullman, 1996).

Matrix participants are temporarily freed from the ego that may judge or guard against the knowledge and insights from dreams and associations. This move away from the ego in social dreaming enables participants to express feelings as shared knowledge, which Gordon Lawrence called the knowledge of the Sphinx as opposed to the repression of the Oedipus (Lawrence, 2003: 268). In the sharing process, the individual dream is transformed into a social dream, thereby affording an emotional distance from the personal concerns of each individual. In therapy, someone may feel guilty for feeling anger towards a parent or intimate friend. In this case, the ego is at play with an individual's sense of self, possibly touching upon issues of infant development that for reasons to be discovered in therapy have not been satisfactorily concluded in adulthood. This is the business of the individual in their relationship with the therapist. In social dreaming, however, the non-judgemental space of the matrix is spontaneously co-created between the participants with the minimal presence of the hosts (facilitators) who specifically reject any therapeutic role. Conditions typical of the therapeutic encounter, such as transference, countertransference, projection, containment and holding, are absent from the matrix. Through the sharing of dreams that co-resonate among participants, the individual becomes 'not me'. This is the opposite of what happens in psychodynamic groups where 'me-ness' may be exacerbated (Hatcher Cano, 1998).

Charlotte Beradt's book *The Third Reich of Dreams* (1968) influenced Gordon Lawrence to devise his social dreaming method by demonstrating the way dream expression could be socially rather than individually relevant. Beradt collected thousands of dreams in the 1930s, prior to Hitler's rise to power. These apparently prophetic dreams show accurately the impact of propaganda in a tyrannical regime. Collective complicity and guilt are recurring themes.

Similarly, the theme of complicity appeared in a matrix focused on the theme of whiteness: A white woman dreams of being on a ship. She didn't witness herself boarding, she cannot see the body of the ship, yet she knows she is sailing. A storm rages at sea and the ship threatens to capsize. In the post-matrix session, she realises this ship is a metaphor for the slave ships. She is white and not in the hold, she inhabits the upper deck or upper stratum of society. The dream is a comment on her complicity, and echoes Bayo Akomolafe's view that as we live with the legacy of slavery, we still inhabit those slave ships (2022). The power of social dreaming in creatively bringing whiteness into contact with the experience of the black slave has been identified and analysed on several occasions, adding weight to the evidence of the way social dreaming works to bring complexity and the impossible to bear upon daily experience (Manley, 2010; Manley and Trustram, 2018). In instances such as these, the impossible becomes not just possible but 'compossible', something that is a 'composite of individual-world-interindividuality' (Deleuze, quoted in Manley, 2020: 436–437). The matrix, layering dreams and associations into a complex collage of experience, liberates participants through different identities and perspectives to imagine from a position of compassion.

In social dreaming, the 'snowflake' seating arrangement, designed – like the couch – to remove eye contact among participants, is one strong feature of the setup. While therapy directly addresses a client's sense of isolation with a problem and supports working through it, social dreaming does not work at the level of ownership of emotional pain and worry. Affects are held in images and belong to the shared scenes that make up the matrix. The potential therapeutic benefit of the social dreaming matrix emerges when a participant, temporarily freed from ego-centred concerns, can access ecological and systemic patterns, belonging to and consisting of a complex wholeness (Gosling and Case, 2013). Each person in the matrix becomes part of this whole. In these ways, social dreaming matrices redraw the boundaries from the individual to a shared, social experience.

In climate psychology, it has become apparent that human egos in the modern world have become separated from nature, treating the natural world as 'Other' (Hollway et al., 2022). Complexity and wholeness require an ecological relation to the natural world, one in which we humans are a part. From an eco-therapeutic perspective, a person's identity does not end at the boundary of their skin. A human is of the earth and in communion with the more-than-human world (Abram, 2011). The non-personal perspective that social dreaming affords means that humans' relations with nature resurface everywhere in the process: the images of water and air are examples, becoming central to the meanings of the matrix once ego-centred relations are retired. Because dreams playfully give dreamers experiences outside their own skin, social dreaming allows a focus on the more-than-human world where there may be an attunement with animals, insects and birds. At other times they may be chasing, biting or stinging. In either case, participants remain curious and engaged rather than fearful. Maybe these other species are conveying important messages:

> *Each night, as a child, I had a recurring dream about being a whale. I sensed my immense body around me, flying gracefully beneath the waves. Roaming the seas, I ate tons of krill each day.*

We can let go of personal identity and, through the experience of dream or association, tap into a shared source or the associative unconscious (Manley, 2018). Lawrence, drawing on Bion's idea of the unconscious, called this 'the infinite', the aspect of self that is connected to all that is (1965 [1984]). This is the wisdom that can be found beyond ego-centred thought. From an ecological perspective we might refer to this complexity as Lovelock's Gaia (Hollway et al., 2022: 8–9).

## Conclusion

This chapter has identified two ways in which Social Dreaming works therapeutically: first, the ability to access complex problems without fragmentation or reduction, and second, the temporary retirement of individual egos from the social dreaming setup. Being able to express complex problems in this way also opens up the potential for engagement through action, something only possible once 'nameless dread' (Bion, 1967 [1984]) is made available for thinking. In this way, a shared social space produces a shared knowing, and engagement can take place beyond individual agency. This contributes to self-awareness and new knowledge that can be preserved – and used – beyond the Social Dreaming Matrix itself.

## Notes

1 Acknowledgement: This chapter is a reworking and development of an article published in the British Psychoanalytic Council Journal, *New Associations* (Manley and Hollway 2020).

2 Ten on-line social dreaming matrices were held weekly, organised by Duke University (USA) and the Centre for Social Dreaming (UK), May–July 2020. The matrices included participants from all over the world and a range of nationalities. Some of the expressions in the transcripts are from speakers for whom English is not the first language. Participant material is anonymised to preserve confidentiality.

3 Methodologically, it is important to note that the analysis of social dreaming data (some would refer to it as 'interpretation') always draws from a wider and more radically interconnected whole than is apparent from the extracts presented in a necessarily linear fashion in the text. The significance of water dreams here is one instance; the same goes for treatment of later extracts.

## References

Abram, D. (2011) *Becoming Animal*. London: Random House.

Akomolafe, B. (2022) The Invisible Constituency of the Slave Ship: The Slave Ship Never Really Disappeared. www.bayoakomolafe.net/post/the-invisible-constituency-of-the-slave-ship [Accessed 20 July 2023].

Beradt, C. (1968) *The Third Reich of Dreams*. London: Quadrangle Books.

Bion, W. (1965 [1984]) *Transformations*. London: Routledge.

Bion, W. (1967 [1984]) *Second Thoughts: Selected Papers on Psychoanalysis*. London: Routledge.

Bollas, C. (1987) *The Shadow of the Object*. London: Free Association.

Freud, S. (Strachey, J. (Trans.). 1991 [1900]) *The Interpretation of Dreams*. London: Penguin.

Gosling, J. and Case, P. (2013). Social Dreaming and Ecocentric Ethics: Sources of Non-rational Insight in the Face of Climate Change Catastrophe. *Organization*, 20(5): 705–721.

Hatcher Cano, D. (1998) Oneness and Me-ness in the *baG*? In: P. Talamo, F. Borgogno and S. A Merciai (eds) *Bion's Legacy to Groups*. London: Karnac, pp. 83–95.

Hollway, W., Hoggett, P., Robertson, C. et al. (2022) *Climate Psychology: A Matter of Life and Death*. Bicester: Phoenix Publishing House.

Lawrence, W. G. (2003) Some Thoughts on Social Dreaming. In: W. G. Lawrence (ed.) *Experiences in Social Dreaming*. London: Karnac, pp. 267–272.

Long, S. and Harney, M. (2013) The Associative Unconscious. In: S. Long (ed.) *Socioanalytic Methods. Discovering the Hidden in Organisations and Social Systems*. London: Karnac.

Manley, J. (2010) The Slavery in the Mind: Inhibition and Exhibition. In: W. G. Lawrence (ed.) *The Creativity of Social Dreaming*. London: Karnac, pp. 41–53.

Manley, J. (2018) *Social Dreaming, Associative Thinking and Intensities of Affect*. London: Palgrave.

Manley, J. (2019) Associative Thinking: A Deleuzian Perspective on Social Dreaming. In: S. Long and J. Manley (eds) *Social Dreaming, Philosophy, Research, Theory and Practice*. London: Routledge, pp. 26–39.

Manley, J. (2020) The Jewel in the Corona: Crisis, the Creativity of Social Dreaming, and Climate Change. *Journal of Social Work Practice*, 34(4): 429–445.

Manley, J. and Trustram, M. (2018) "Such Endings That Are Not Over": The Slave Trade, Social Dreaming and Affect in a Museum, *Psychoanalysis, Culture and Society*, 23(1): 77–96.

Manley, J. and Hollway, W. (2019) Climate Change, Social Dreaming and Art: Thinking the Unthinkable. In: P. Hoggett (ed.) *Climate Psychology: On Indifference to Disaster*. London: Palgrave, pp. 129–152.

Manley, J. and Hollway, W. (2020) Breaking the Shell, Rescuing the Seed. *New Associations, Journal of the British Psychoanalytic Council*, (32): 1–2.

Morton, T. (2013) *Hyperobjects*. London: University of Minnesota Press.

Serra Undurraga, J. K. A. (2023) What Are Our Psychotherapeutic Theories and Practices Producing? *European Journal of Psychotherapy & Counselling*, 24(4): 1–16.

Ullman, M. (1996) *Appreciating Dreams: A Group Approach*. New York: Cosimo-on-Demand.

Winnicott, D. W. (1991) *Playing and Reality*. London: Routledge.

Chapter 19

# 'Ways of Being' When Facing Difficult Truths: Exploring the Contribution of Climate Cafés to Climate Crisis Awareness

*Gillian Broad*

### Take a Moment ...

Imagine you are in a warm and hospitable space alongside eight other people, all sitting around a large table covered with mugs of tea or coffee and a large cake at the centre. As the host welcomes everyone to this climate café they invite the people present to consider what they might need in order for this to be a safe and supportive space for them to engage in a conversation about the climate crisis and their emotional responses to it ... and that is how a climate café begins.

### The Background to Climate Cafés

As it becomes more evident that climate and ecological breakdown are a clear and present danger to our safety and wellbeing, we all increasingly need to talk about what our changing world means for us in terms of impacts at the personal, family and societal level. Adapted from the Death Café model (Underwood, 2014), Climate Psychology Alliance (CPA) began running climate cafés in 2019. A climate café is a simple, hospitable, empathic space where fears and uncertainties about our climate and ecological crisis can be safely expressed. It offers an accessible and supportive emotional approach, outside of the consulting room. The chapter begins with an overview of the rationale and theoretical underpinning of climate cafés and moves to focus on the 'ways of being' established at the outset of a café that set the tone and purpose of this unique and paradoxical space.

### What Is the Purpose of a Climate Café?

There are different versions of climate cafés. Those run under the auspices of CPA are positioned as spaces to share our responses to climate change, and specifically to offer a feeling and thinking space and a haven from busy-ness and activity. It is not that climate action is disapproved of, rather that this

DOI: 10.4324/9781003436096-36

space is action free, allowing the affective dimensions of the climate crisis to come to the fore.

Climate cafés are difficult to 'pin down' in terms of a descriptor that accurately captures what they are there to do. While not therapy groups, they offer a hybrid liminal space that has a broadly therapeutic intention. Within this therapeutic atmosphere there is no expectation that people should commit to attend beyond the café they join and no risk of them feeling they have been inveigled into some sort of exclusive club. The invitation to attend is open to all.

## What Happens In a Climate Café?

Through the ancient tradition of hospitality, and its 'gift economy' approach, a climate café seeks to make individuals more aware of their sense of mutual dependence, care, intimacy and creatureliness. Climate cafés, then, are open, respectful and confidential spaces for people to gather and express their views safely. There is no intention through the course of a café of participants 'being led' to any conclusion or particular course of action. Cafés are open to anyone and are particularly geared to respond to the needs of people – concerned citizens – who have begun to experience difficult feelings about the climate crisis and who do not have supportive networks or spaces where they feel able to talk and share their feelings. Equally climate cafés can be a recuperative space for activists, who often do not have the opportunity to acknowledge their feelings within their activism spaces.

In terms of what actually happens in a climate café, the Climate Psychology Alliance version has a particular format. The café begins with attendees being welcomed and invited to select a natural object from a bowl placed on the table, containing a selection of natural objects, including leaves, twigs, feathers, seedheads, stones, flints, shells and lichen. (Online versions invite participants to bring an object to mind or pick up one if they have something near to hand.) Once everyone has selected an object from the bowl the group members take it in turns to introduce themselves in relation to how what they have selected connects them to the climate crisis. The exercise begins to help people familiarise themselves with the purpose of climate cafés – to talk about difficult feelings in relation to the climate crisis. Everyone takes a turn to speak and hears their voice in the room. Once the opening round is completed, the second phase provides an opportunity for group members to connect to the different stories they have heard. In the second stage of free-flowing discussion, if an attendee chooses not to speak, that is their prerogative.

## Establishing 'Ways of Being' In a Climate Café

In order to realise the intentions underpinning the climate café model, there is a need to pay careful attention to what in formal therapy contexts might be referred to as the necessary conditions to establish the 'therapeutic milieu' in this hybrid, liminal space.

Acknowledging this dilemma gets to the heart of the paradox of the climate café space. In order to fully engage with the extent and implications of the climate crisis, we need to imagine it in some detail so as to be able to explore our complex feelings and thoughts, which may be hard to talk about. Kari Norgaard (2011) in her study of a Norwegian skiing town refers to the *socially constructed silence* or social taboo that developed as a result of the unbearable contradictions inherent in the impact of the climate crisis on us all. Climate cafés are intended to enable people to begin to share their feelings in order to think and talk about them and reduce our unhealthy dependence on defensive strategies as a coping mechanism. With sturdy enough support structures in place, most people can sustain challenging feelings without resorting to dissociation, numbness or blind panic. They can engage with difficult truths while staying connected and grounded. A climate café aims to be such a structure – a container that is strong enough to allow the feelings of fear and anxiety, and other emotions such as anger, helplessness, sadness, grief or depression, to be explored.

A climate café *is,* then, a low key, friendly and hospitable opportunity to discuss the difficult truths and express painful and hard to bear feelings that are becoming more and more recognised in relation to the climate crisis and our emotional responses to it. Simultaneously, on account of its emotionally oriented focus on the climate crisis, it is also a potentially challenging and discombobulating space. These contradictions, or paradoxes, in relation to structure and impact, seem to permeate the climate café format. The first such contradiction arises at the beginning of the *informal* café space when attendees are invited to engage in what might be considered to be quite a *formal* activity – a discussion to determine the café's 'ways of being'. How then, in light of these potentially contradictory and polarising group dynamics, is it possible for a safe and supportive atmosphere to be created in a climate café?

## Exploring 'Ways of Being' In Climate Cafés

It is only very recently that we have renamed what we originally referred to as 'ground rules' for climate cafés to 'ways of being' in climate cafés. This shift in terminology arose out a realisation that 'ground rules' had a technical and formal ambience that was not in keeping with the ethos of the climate café space. Nonetheless, in order for the café spaces to be experienced as safe, some boundary setting is unavoidable. Adopting this more nuanced and affective language of 'ways of being' speaks to the heart of the café experience, which seeks to create opportunities for individuals to express heartfelt feelings that require respectful conditions for that to be possible.

The manner by which we have developed the 'ways of being' conversation has also developed over time, alongside, although not entirely co-synchronously with, the change in terminology. Our earlier approach was to begin the café by outlining the pre-determined, at that point in time, 'ground rules' that we, as the café hosts, had identified as the key and necessary ones that would allow the group to function in a safe and supportive manner. As our practice has developed, we have adapted our approach and brought the identification of these 'ground rules' into the café space itself, inviting the attendees to take ownership of what is important for them. Interestingly, the issues that have been highlighted in this revised approach have broadly aligned with those that were previously pre-determined. The more inclusive stance to establishing them, however, feels more collaborative and most recently this has contributed to the linguistic shift to using the term 'ways of being', underlining our shared responsibility for how we are together in this unique space. Each of the ways of being that are most commonly acknowledged at the beginning of a climate café are explored below.

### *Confidentiality*[1]

Perhaps the most familiar of all the 'ways of being' for any space where people are making themselves to some extent vulnerable, is a commitment to confidentiality. While the broad focus of the café conversations can be referred to outside of the group (and such conversations are important to spread the word about the value of the café experience), specific content introduced by individuals within the café must *not* be shared outside of the group. Such a commitment is essential if an emotionally containing space is to be created that enables individuals to be vulnerable with others and to express and share their climate-related anxieties.

### *All Emotions Are Welcome*

As acknowledged above, climate anxiety manifests in a plethora of emotions. These powerful feelings can be experienced as unbearable. The size and uncertainty of the problem are too great for us to process; in some instances, this can generate feelings of rage and terror in response to the fear of annihilation and collapse. During the 'ways of being' phase of the café we underline that all emotions are welcome. When the café begins, it is common practice for one of the two facilitators to begin the opening round with reference to their chosen natural object, as outlined above, and to share how it connects to their relationship with the climate crisis perhaps alluding to feelings of sadness and despair. On other occasions it might reflect anger and frustration. Whatever the emotion expressed, it is not uncommon for this suite of emotions to become the dominant discourse in the café as people make connections to the contributions already made. One way we ensure that everyone feels their emotional responses are valid is to reiterate this specific 'way of being' (that all emotions are welcome), as and when appropriate, throughout the duration of the café. Specifically, towards the end of the café, it is our practice to intentionally acknowledge that there might be feelings people have not yet expressed that they would like to speak to before the café finishes. Often this invitation is taken up by someone and the emotions that might have been, up until that point, absent from the conversation are identified and owned.

In a recent café, it was noticeable that having begun the opening round in my role as the facilitator, I shared some difficult feelings of guilt I was experiencing in relation to some travel choices I had recently taken. This led to the café attendees following on from this with a strong focus on feelings of blame, shame and guilt. Realising the direction that the café had taken led my co-facilitator, in the final section of the café, to astutely acknowledge that there may be other feelings people wanted to bring into the space. This invitation elicited an immediate response from someone who identified how angry they were feeling about what was happening in the world.

Given the widely accepted idea of the socially constructed silence (Norgaard, 2011) that surrounds the climate crisis, it feels important that this acknowledgement that 'all emotions are welcome' is an established part of the café process.

### *Respecting, Empathising and Not 'Fixing' Feelings*

Building on the invitation to express emotions outlined above, in the course of a café it is not uncommon for individuals to have embodied emotional responses – tears being the most common. Unsurprisingly, other café members are often quick to empathise with the distressed person. In some instances, however, there can be a defensive inclination to 'shut the feeling down' and to offer false reassurance, such as 'it might not be so bad', or that there are more grounds for hope than have been suggested – the overly positive position mentioned earlier.

On one occasion a woman expressed her deep ambivalence about whether she felt able to have children. Such deeply personal admissions always represent profoundly moving moments in a café. Another woman who had already identified as a mother immediately sought to offer reassurance by explaining how she had come to terms with similar feelings by committing to bring up her daughter as a responsible, environmentally aware citizen. In the moment, my co-facilitator and I did not recognise quickly enough how this well-intentioned response was denying the first woman's feelings. As a consequence, after the café had finished this woman contacted my co-facilitator and expressed her distress at being, in effect, silenced. When such responses to powerfully expressed feelings arise, it is the responsibility of the café facilitators to notice the dynamic and remind the group that all feelings are welcome and that it is not the responsibility of group members to offer reassurance or seek to reduce/remove the pain, rather their responsibility is to empathise and bear it together. The role of the café members is to listen with interest and empathy to each other's views, without imposing their own. It is *not* the role of café members 'to fix' each other's feelings.

### *Turn-taking and Taking Up Space*

The climate café space is intended to be inclusive and respectful. From the outset, as we establish our 'ways of being' together, the practice of turn-taking and allowing space for everyone to speak is usually identified by a café member as an important one and, if not, a facilitator will ensure it is raised. What this 'way of being' highlights is the importance of everyone having equal opportunity to speak and for the group members, including facilitators, to be mindful of how much space they are taking up. The café conversation is intended to be spontaneous and not structured (beyond the pattern of the two rounds), therefore, people are reminded to try not to interrupt or speak over each other. Perhaps unsurprisingly the 'no action' aspect of the café often collides with this 'way of being'. When, on occasion, someone needs to prove a point or persuade someone of the validity of a particular view, it invariably involves them taking up too much space. As with all instances of the agreed 'ways of being' being disregarded, it is the role of the facilitators to gently bring this to the attention of the group and invite people to stay focused on what they *feel*, as opposed to what they are thinking or what they want to 'do'.

### *Sitting In and with Silence*

Although 'normal' cafés are usually places for conversation, and in fact if the climate café is held in an actual public café there will probably be background hubbub, a distinctive feature of climate cafés is their ability to tolerate silence. That said, silence is an unfamiliar 'way of being' for many people, who may feel the need to fill it. Making clear that silences in the café are acceptable and are often a sign of deep affect at work assisting the processing of difficult feelings aroused, helps ease any anxiety or discomfort people unfamiliar with silence might have in collective spaces. Facilitators might even consider it helpful for the group to deliberately take a moment to pause and reflect silently on what has gone on in the café.

### *Acknowledging 'Ouches'*

Perhaps the ground rule that is most difficult to articulate is what we have come to call 'ouches'. Ouches, a term which originates from The Work That Reconnects (Macy and Johnstone, 2012), refers to potential moments in a café when something might be said that another member of the café experiences as hurtful. Often, they are unconsciously enacted and similar to micro-aggressions in relation to racism; they need to be made visible in order to highlight our complicity in oppressive structures and our hurtful behaviours. Unless the group feels it is important that they address the issue that has provoked the 'ouch' further, it is not necessary for anything more to be done with these 'ouch' moments, other than acknowledgement; identification is enough.

## What Do the 'Ways of Being' Achieve?

So why do we place such weight on establishing these 'ways of being' at the outset of a café? As with any group endeavour, creating boundaries around the group activity is recognised as necessary for the group to feel contained and safe enough to stay focused on the café's intended purpose. Although an informal social space, not a formal working group, it does have an explicit purpose beyond socialising, hence 'ways of being' create the group ethos and focus. Importantly, taking the time at the outset of the café to agree the ways of being with the group establishes a shared sense of commitment to the space and a feeling of being connected. This building of a group identity is critical for the group to feel safe enough to open up and for the space to be one where individuals can feel confident in expressing their vulnerability.

Since establishing this way of working and paying attention to the 'ways of being' as café facilitators, we have become increasingly aware of how quickly and powerfully people can come to feel connected in the café space. Both in online and in-person cafés people speak of feeling a bond across their shared expressions of climate grief, anger, rage, despair and sadness. It appears that the group medium creates a strong sense of solidarity and diminishes the sense of isolation and loneliness, which the social silence that has been constructed around the climate crisis can elicit.

Café attendees frequently refer to a sense of relief that they can share their climate distress and dismay with others who understand it and are not afraid to express and explore these difficult feelings. One way in which a café can be concluded is to invite each person to say one word to describe how they are feeling at the end of the café. More often than not at least one person, and frequently more than one, will choose the word 'connected', and as a café facilitator I often find that is the word that springs to mind (to heart?) for me. Another word I have found myself using, not in my usual vocabulary, is 'solace' – finding comfort or consolation in a time of distress or sadness. It feels deeply rooted, powerful and apposite.

The role of the 'ways of being' in creating this sense of connection and containment can be partly understood in terms of how they help to structure the way the group behaves. To understand how this works it is helpful to refer to our own experience as facilitators of convening climate café facilitator workshops, to equip prospective facilitators with the relevant knowledge and skills for the role. In the course of the workshop, we invite attendees to raise any questions they have about the role and concerns regarding what might happen in the café space. Almost without fail the following two concerns are raised: how to manage dominant people in the café and how to deal with powerfully expressed emotions, particularly anger and sadness. Responding to these articulated concerns has reinforced for us the importance of clearly establishing the 'ways of being' from the outset. With these behavioural expectations and boundaries in place, it is always possible to respond to the prospective facilitators' concerns about managing a café by reminding them of the 'ways of being' that they will have agreed with their group at the outset. They act as safeguards if it feels the dynamics in the café are taking it 'off task'. For example, if someone dominates the conversation and does not allow contributions from others, the facilitator can gently interject, reminding everyone of the 'ways of being' that recognise it is important for the café space to be one in which everyone has equal opportunity to contribute, if they wish to.

As the café ends it is our practice to make it known that the facilitators will remain available at the end of an in-person café for a while, should a group member wish to speak with them individually. In the case of online cafés, facilitators provide their contact details so that attendees can contact them after the café should they feel the need. The powerful impact of climate café conversation should not be underestimated; the careful attention paid to how cafés are set up and ended underlines this. Our expertise in hosting cafés across the United Kingdom is becoming increasingly honed as we incorporate our learning along the way into 'our ways of being'. That said, we are mindful that 'ways of being' reflect a dynamic state; we will never arrive and there is no room for complacency. The climate crisis is all too real, and we need to stay on our toes ready to respond in whatever ways are required to ensure climate cafés remain focused and able to offer the solace we have described.

## Note

1 In the spirit of this, any climate café participant material in this chapter is anonymised to preserve confidentiality

## References

Macy, J. and Johnstone, C. (2012). *Active Hope: How to Face the Mess We're in Without Going Crazy*. Novato, CA: New World Library.

Norgaard, K. (2011) *Living in Denial: Climate Change, Emotions and Everyday Life*. Cambridge, MA: MIT Press.

Underwood, J. (2014) *The Death Cafe* [Online]. Available at: https://deathcafe.com/what/ [Accessed 9 May 2019].

Chapter 20

# The 'Ticking Clock Thing': Climate Trauma in Organisations[1]

*Rebecca Nestor*

## Introduction

Working on climate change affects people significantly. As set out in chapter two, the climate crisis evokes a range of powerful unconscious responses and their associated defences, and people working on climate are particularly exposed to these impacts. The task involves facing what is sometimes described as the 'super-wicked' problem of the climate and ecological crisis and the specific emotional effects of this are described below.

Super-wicked problems are those that will get harder and harder to address the longer we leave them, but which are also subject to the human tendency towards irrational discounting of the future. Those working on super-wicked problems know that they do not have the luxury of time, so feelings of urgency and enormous frustration will be common. Those who are in the best position to address a super-wicked problem are those who caused it, who are also those with the least immediate incentive to act within the necessary timeframe. So, confrontation with the painful realities of power and structural inequalities is inevitable. And a super-wicked problem suffers from the absence of any global authority (legal or regulatory) that matches the scope of the problem (Lazarus, 2008: 1160–1161). So, collaboration is essential, and it is likely that feelings of vulnerability will be present because of the absence of authority. In other words, super-wicked problems require almost superhuman levels of collaboration and sophistication, but their pressures mean that those working on them are more likely to be stressed, angry and vulnerable.

Climate work, then, involves working collaboratively between contradictory contexts on an existentially important, urgent, frightening and near-impossible task. Studies of organisational life focusing on sustainability professionals and others working on climate in organisations, or communicating it to the public, unsurprisingly indicate that this work carries significant emotional demands. There is a heavy burden of responsibility and a requirement to work within 'business as usual' and be convincing there, while facing the reality of the climate crisis and envisaging a different future in the face of personal terror,

DOI: 10.4324/9781003436096-37

collective trauma and systemic barriers to change. Psychological and social defences are to be expected in this challenging work.

This chapter explores the particular emotional 'feel' of climate work, drawing on a range of studies including my own doctoral study (Nestor, 2022), which convened a group of climate communications leaders in a variety of roles and organisations: the head of climate campaigning for a large international charity; two heads of small climate charities, one focusing on communication skills and the other on lifestyle change; the sustainability lead for a large religious organisation; a local authority officer; and the chair of a local climate action group. The group held seven meetings during most of 2019, online and in person. In addition to the group meetings, I conducted individual interviews in person as part of the triangulation necessary for connecting group experience to the wider dynamics of members' work, and a reflective note was produced for members to consider. I also made very active use of my own emotional and behavioural responses and dreams.

## The Emotional Impact of Climate Work

### *Threat, Vulnerability and Fragility*

People working on climate are vulnerable to external threat. Paul Hoggett and Rosemary Randall conducted a comparative study of the working cultures of climate scientists (primarily those engaged in public communication of climate change) and climate activists (Randall and Hoggett, 2019). Their study found that some scientists' emotional challenges derived more from the battles with colleagues about how the science should be presented to the public, and the reception from the press and media when they tried to explain the implications of the science, than from their fear of the effects of climate change itself. As one respondent said, 'Climate scientists have now become afraid of any sort of situation in which they feel there is a dispute, and they're trying to avoid that at all costs … they don't want to have a conflict with their opponents' (Respondent quoted in Randall and Hoggett, 2019: 250). Climate activists have their own – perhaps more visceral and physically threatening – version of this experience from their exposure to police and other representatives of the state:

> when the police have come round and they've fucking walked on our bed in their boots and emptied all the cupboards and drawers out, to no real effect, you know, it's just what they do – that's traumatic, and the neighbours seeing the police pouring into the house and things …
>
> (Respondent quoted in Randall and Hoggett, 2019: 248)

My doctoral research uncovered a sense in climate work of being fragile, on the edge, or in a dangerous position nearing a tipping point or approaching collapse – perhaps mirroring scientific realities. One respondent was very explicit about this feeling:

> We had a meeting last summer and we, we spent two days talking about, really, how do we cope with being at the edge, on the edge of social collapse, societal collapse, and this line from Rumi … which is: sit still and listen, for you are drunk and we're on the edge of the roof.
>
> (Respondent quoted in Nestor, 2022)

Other respondents reported feelings at work of having nothing to fall back on, and feeling preoccupied with what others were feeling; two respondents reported obsessive worrying. These personal anxieties and vulnerabilities echo the global lack of authority in a position to address the crisis and the huge uncertainties in the work. Is the (social) science right? How do we know what will be effective in public engagement? Will the technology work or will it fail? The feeling is of a precarious coping, strongly motivated but at risk of being destroyed from within and without by unmanageable forces; an awareness of the potential for emerging disorder, combined with a sense of not knowing and of being uncertain what is to be trusted.

### *Contradictions and Tensions*

Wright, Nyberg and Grant (2012) conducted an interview study of 36 environmental sustainability professionals in corporate settings. Engagingly titled 'Hippies on the third floor', the study explored how their respondents managed the felt contradictions in their experience. As the authors point out: 'Climate change threatens not just our economic and social way of life, but our very sense of who we are as individuals' (ibid.: 1453), and as change agents working within the paradigm that shareholder value and profit are what count, the study's respondents are located at the heart of this contradiction. The respondents managed the challenges by switching between different identities depending on their audience and their own emotional needs, creating a sense of coherence through narratives in which they featured as heroes. Their findings are supported by other studies such as Penny Walker's practitioner perspective on the emotional challenges faced by sustainability professionals (Walker, 2008a, 2008b) and Nadine Andrews' insightful PhD and journal article inquiring into the experience of six sustainability managers and leaders in the UK and Canada. Andrews identifies the psychosocial factors affecting her respondents' work (threats and tensions, coping methods, contextual influences on the efficacy of the coping methods, and the ways in which all of these relate to each other in a feedback loop). Like Wright and colleagues, she finds the use of heroic narratives and she also identifies a focus on positivity as a coping strategy (Andrews, 2017).

My doctoral study (Nestor, 2022) found tensions and contradictions in climate communications work. For example, there was a belief that the external messaging and the needs of others required positivity, while internally there were feelings of despair and exhaustion that could not be attended to. This clearly creates an emotional strain, with one respondent making an evocative comparison with the experience of 'people who work at Disneyland, who have to be relentlessly positive, and there was some research done recently … it found that they were the most grumpiest, miserablest people at home' (Respondent quoted in Nestor, 2022). Other contradictions in my study included an acute awareness of the urgency of the climate crisis, and a wish to reach commitment and move fast, while knowing that the work of engagement requires relationality and patience. Some felt a sense of there being an almost infinite amount to do while feeling exhausted and drained. The sense of contradiction extended into experiences of the organisation itself: for example, respondents described both a love for the organisation's values and a hatred of its necessary bureaucratic processes ('I'm filling in spreadsheets,' complained one member, 'to solve climate change').

### *Social Defences in Climate Work*

In the previous section we saw the emotional impact of climate work. How might organisations respond to this impact?

Engagement in climate work can help with difficult feelings through creating a sense of agency and contribution; but in organisations that do climate work, climate emotions are likely to become intensified through connection with the work. It is likely then that such organisations will experience what Isabel Menzies Lyth (1960) termed 'social defences' against unmanageable feelings: for example, the use of splitting,[2] projection,[3] and projective identification[4] to create polarisations and stereotypes, which become embedded in the organisation's ways of working and its structures. An organisation struggling with social defences might, in order to avoid the pain of its primary task, create work that does not serve the purpose of that task (sometimes known as 'anti-task' behaviour), and it would hardly be surprising to find that an organisation whose primary task required its members to face the climate crisis would seek ways to avoid the pain of this. Some of the experiences described above have a polarised feel to them, and similar splitting was found in Peliwe Mnguni's (2010, 2008) exploration of a partnership for environmental sustainability in Australia. Mnguni identifies splitting between those parts of the partnership representing 'close-to-nature' and those representing 'estrangement from nature' (ibid.: 122); the claiming of all reparative intent for one's own organisation and projection of all responsibility for 'socio-ecological degradation' into other organisations (ibid.: 123); and the idealisation of working on the land compared with working in the town. She also identifies organisational social defences in the creation of and collusion with meaningless tasks and targets. Social defences in organisations can involve the unconscious creation of systemic or structural barriers to working effectively together: an investigation of civil servants in the US, Australia and Canada (Temby et al., 2016) found that the research subjects thought that collaboration was crucially important when working on climate change adaptation, but that they experienced systemic barriers to its successful operation. A 2012 study of the difficult relationships between UK environmental voluntary groups and local government found that the voluntary groups were cynical and suspicious of local authorities' motives for partnership, and more comfortable working with each other than with local authorities. However, they were sometimes prepared to be pragmatic and responsive if it would help to build a shared power base.

## Leadership in Climate Work

Leadership in climate work is particularly challenging, especially when we consider climate change as a super-wicked problem. In much leadership discourse, leaders in climate and sustainability work are described as needing the following rather alarming list of qualities: the ability to work across organisational and disciplinary boundaries for multi-party collaboration; communication and influence; empowerment; ethics, responsibility, respect for the natural world; innovation; long-term thinking and purpose; maintaining many strong relationships; self-awareness, self-reflexivity and critical thinking; narrative skills; systems thinking; tenacity and resilience; and an understanding of power in social relations (Quinn and Baltes, 2007; Parkin, 2010; Mead, 2014; Grayson, 2017; Cambridge Institute for Sustainability Leadership, 2021). This emphasis on the value and significance of collaboration, boundary-spanning and handling uncertainty sometimes looks as though a significant development in human cognitive capacity is expected. But the reality of the lived experience of leaders from the empirical studies described in this chapter seems often to be one of huge difficulty with collaboration in the most ordinary sense of the word, unsurprising in the context of threat and trauma. Climate leadership may feel suffused with failure, disappointment and cynicism, and with mutual suspicion and limited resources creating difficulties in collaboration. Jonathan Gosling defines the challenge for leaders of how to hold on to hope and some idea that things could be done right, while still acknowledging the likelihood of failure at a 'point at which our own meaning systems are stressed to the point of breakdown' (Gosling, 2017).

## Trauma in Climate Work

The quotation from Jonathan Gosling which closes the previous section hints at the possibility in climate work of the breakdown of meaning associated with trauma – and indeed, the work of people involved with climate change can be seen as potentially traumatic in several ways. They are exposed to an awareness of natural catastrophes (for example, Pihkala (2019) describes environmental researchers as experiencing secondary trauma as a result of their work). They are attacked by an external group (as Randall and Hoggett (2019) describe scientists' and activists' experience). They repeatedly experience being overburdened by the demands of this super-wicked problem that lie beyond their capacity to cope. They may feel the 'moral injury' and sometimes unbearable pressure of being associated with organisations which are part of the establishment that is resisting the necessary change. They may be particularly exposed to collective climate trauma (Woodbury, 2019; Bednarek, 2021), because their work requires them to be in touch with the collective challenge. Some aspects of climate communications work – working across boundaries, communicating science to non-scientists, working within the very 'business as usual' that is destroying our chances of survival – intensify the cognitive impossibilities of contemplating the prospect of human intervention in the climate we live in. The work of engaging and communicating calls for steps to be taken that are denied and unresourced within the social systems we have, and this brings participants face to face with the limitations of the social systems within which their work takes place. These are not surface or cognitive difficulties. They involve intensely frightening and desperate feelings.

## How Can Therapeutic Practitioners Help?

What support can be offered to those experiencing these dynamics?

Paying attention to the body to heal the disconnect with our creaturely selves may be an important aspect of support for collective climate trauma, as it is known to be for collective racial trauma (Menakem, 2021).

Access to nature, as part of a conscious intention to build the knowledge of connection to and dependence on the other-than-human, has been shown to have a healing effect, and when combined with an openness to sharing difficult feelings it can boost agency and what Sally Weintrobe calls 'lively entitlement' (Weintrobe, 2021) in people working on climate change. As Jo Hamilton puts it, 'making space to explore how emotions influence thoughts, mindsets and behaviours … acknowledging or expressing difficult or painful emotions can support active engagement' (Hamilton, 2020: 112).

Organisations in this field may need to be open to people needing regular time out – both to access support for healing and because every so often people may need to experience the temporary 'necessary derangement' that Steffi Bednarek (2020) proposes as part of our collective recovery from the permanent collective derangement we are living in. Such time out might support the development of the new capacities seen by some as necessary for climate change leadership: thinking outside the human frame and collaborating across species boundaries. Regular, accepted absences might also encourage mutual dependence and sharing of leadership tasks, reducing the risk of overload on individuals.

However, climate trauma perhaps carries a particular kind of resistance to healing. Traumatised individuals and groups commonly feel in their minds and bodies as if the original trauma is being repeated, and in addition to this, in climate trauma there is the fact that the originating attack is in fact objectively still happening – all around us and in us as we live our unavoidably fossil-fuelled lives. Recovery and healing must necessarily then be a slow, hesitant, sometimes impossible process.

Nevertheless, some steps can be taken. Individual therapeutic support from climate-aware therapists could be made more widely available to enable those working with the climate crisis to understand how their early internal experiences may be re-activated by the external trauma of the climate crisis, and to rebuild their connections with their internal 'good objects', as Melanie Klein conceptualised it (Bott Spillius et al., 2011).

In addition to and alongside individual support, it is particularly important that organisational consultants working with climate include group spaces in their offer, for two reasons. First, so that the wider organisational dynamics can be re-experienced, held safely and re-processed in the smaller, facilitated group – something that cannot realistically be achieved in individual work. Second, so that the individual heroic narratives can begin to be replaced with a deeper knowledge, through experience, of the importance of collaboration and interdependence in climate work.

There is consensus that working with trauma begins by people being supported to put their experience of the trauma into words in a contained space, in order to grieve; that coping mechanisms must be validated, because defences are a necessary if not sufficient aspect of responding to trauma; and that body work and working with the more-than-human are core to enabling reconnection and reintegration. For climate trauma Bednarek (2021) proposes a focus on the collective, including the non-human; practices to re-indigenise the mind, helping a return to the knowledge of interdependence that has been denied; and being prepared for shame arising from the experience of beginning to be heard.

Garland (1998) proposes that a group whose trauma arises from their shared work benefits from interventions that enable them to recreate the meaning of their primary task, because this is 'their most powerful container' (Garland, 1998: 188; see also Bell and Bugge, 2011). Recreating meaning is particularly complex for people working on climate, where the fundamental nature of the work is making meaning out of complexity and uncertainty, where it may feel as though whole structures of meaning have been lost, and where urgency, anger and vulnerability are so prevalent. A group intervention of this kind would need to offer a space where slowing down is supported, where action and decisions are excluded and where anger and fear can be expressed.

The work of sharing experiences of trauma would need to build up to an exploration of how the shared experience can re-shape the organisational members' approach to their work. A therapeutic group like this, its therapeutic task completed or at least ended, might morph into a group whose task is an organisational learning one, supporting the wider organisation in re-thinking how it carries out its primary task in these times of continuing trauma.

## Notes

1 The warmest of thanks to the participants in my doctoral study for their permission to be quoted. The phrase 'ticking clock thing' in the thesis title originated in a statement in one of our meetings, and for me encapsulates the emotional experience of working in climate change communication and engagement: 'we're always working with this like ticking clock thing, ticking clock thing, and we ... we try and go faster and faster and faster and get there quicker and quicker and quicker because of the ticking clock thing'. Thank you all for managing to keep the ticking clock enough in the background to take part in the study; thank you for sharing your feelings and responses, even when these were painful or bewildering; thank you for the work that you do.
2 Splitting: an early defensive response to overwhelm, in which the mind cannot cope with complexity or ambivalence and instead creates binary opposites, 'good' and 'bad'.
3 Projection: when faced with aspects of the self that cannot be borne or are socially unacceptable, individuals may need to deny the presence of these aspects and claim that they instead belong in others.
4 Projective identification: a form of projection described by Melanie Klein, in which the projected aspect of the self is accepted and absorbed by the other.

## References


Andrews, N. (2017) *Psychosocial Factors Affecting Enactment of Pro-environmental Values by Individuals in Their Work to Influence Organisational Practices.* PhD, Lancaster University.

Bednarek, S. (2021) Climate Change, Fragmentation and Collective Trauma. Bridging the Divided Stories We Live By. *Journal of Social Work Practice*, 35: 5–17.

Bell, J. and Bugge, R. G. (2011) Recovery from Crisis in Organizations (With or Without Post-traumatic Stress). *International Society for the Psychoanalytic Study of Organisations Annual Symposium.* Melbourne.

Bott Spillius, E., Milton, J., Garvey, P. et al. (2011) *The New Dictionary of Kleinian Thought.* London: Taylor & Francis.

British Gestalt Journal (2020) *Steffi Bednarek on Necessary Derangement.* Available at: www.britishgestaltjournal.com/features/2020/10/21/steffi-bednarek-on-necessary-derangement-at-the-upcoming-bgj-seminar-day [Accessed 31 August 2021].

Cambridge Institute for Sustainability Leadership (2021). *Our Position on Leadership for the Future* [Online]. Cambridge: Cambridge Institute for Sustainability Leadership. Available at: https://www.cisl.cam.ac.uk/resources/cisl-frameworks/leadership-hub/our-position-on-leadership-for-the-future [Accessed 10 August 2021].

Garland, C. (1998) *Understanding Trauma: A Psychoanalytic Approach*. London: Duckworth.

Gosling, J. (2017) *Purpose, Motivation and Trust: The Elusive Trinity of Leadership. Eric Miller Memorial Lecture, 25 March. Royal College of Obstetricians and Gynaecologists*, London: OPUS.

Grayson, D. (2017) *New Leadership Competencies for Corporate Sustainability*. Available at: www.cranfield.ac.uk/som/thought-leadership-list/new-leadership-competencies-for-corporate-sustainability [Accessed 10 August 2021].

Hamilton, J. (2020) *Emotional Methodologies for Climate Change Engagement: Towards an Understanding of Emotion in Civil Society Organisation (CSO) – Public Engagements in the UK*. PhD, University of Reading.

Kythreotis, A. (2012) Autonomous and Pragmatic Governance Networks: Environmental Leadership and Strategies of Local Voluntary and Community Sector Organizations in the United Kingdom. In: D. R. Gallagher (ed.) *Environmental Leadership: A Reference Handbook*. Thousand Oaks: Sage.

Lazarus, R. J. (2008) Super Wicked Problems and Climate Change: Restraining the Present to Liberate the Future. *Cornell Law Review*, 94: 1153.

Mead, G. M. (2014) *Telling the Story: The Heart and Soul of Successful Leadership*. Chichester: John Wiley.

Menakem, R. (2021) *My Grandmother's Hands: Racialized Trauma and the Pathway to Mending Our Hearts and Bodies*. UK: Penguin.

Menzies Lyth, I. (1960) A Case Study in the Functioning of Social Systems as a Defence Against Anxiety: A Report on a Study of the Nursing Service of a General Hospital. *Human Relations*, 13: 95–121.

Mnguni, P. P. (2008) *Mutuality, Reciprocity and Mature Relatedness: A Psychodynamic Perspective on Sustainability*. PhD, Swinburne University of Technology.

Mnguni, P. P. (2010) Anxiety and Defence in Sustainability. *Psychoanalysis, Culture & Society*, 15: 117–135.

Nestor, R. (2022) *'The ticking clock thing': A Systems-psychodynamic Exploration of Leadership in UK Organisations that Engage the Public on Climate Change*. Professional Doctorate, Consultation and the Organisation, University of Essex and Tavistock and Portman NHS Foundation Trust.

Parkin, S. (2010) *The Positive Deviant: Sustainability Leadership in a Perverse World*. Abingdon: Earthscan.

Pihkala, P. (2019) The Cost of Bearing Witness to the Environmental Crisis: Vicarious Traumatization and Dealing with Secondary Traumatic Stress Among Environmental Researchers. *Social Epistemology*, 34: 86–100.

Quinn, L. and Baltes, J. (2007) *Leadership and the Triple Bottom Line: Bringing Sustainability and Corporate Social Responsibility to Life*. Greensboro, NC: Center for Creative Leadership.

Randall, R. and Hoggett, P. (2019) Engaging with Climate Change: Comparing the Cultures of Science and Activism. In: P. Hoggett (ed.) *Climate Psychology: On Indifference to Disaster*. Cham, Switzerland: Palgrave MacMillan.

Temby, O., Sandall, J., Cooksey, R. et al. (2016) How Do Civil Servants View the Importance of Collaboration and Scientific Knowledge for Climate Change Adaptation? *Australasian Journal of Environmental Management*, 23: 5–20.

Walker, P. (2008a) Organisational Leader or Part of a Wider Change Movement? How Sustainable Development Change Agents See Themselves. *EABIS Colloquium*.
Walker, P. (2008b) Supporting the change agents: Keeping ourselves effective on the journey of change. *Greener Management International*, 54, p. 9.
Weintrobe, S. (2021) *Psychological Roots of the Climate Crisis: Neoliberal Exceptionalism and the Culture of Uncare*. London: Bloomsbury.
Woodbury, Z. (2019) Climate Trauma: Toward a New Taxonomy of Trauma. *Ecopsychology*, 11: 1–8.
Wright, C., Nyberg, D. and Grant, D. (2012) 'Hippies on the Third Floor': Climate Change, Narrative Identity and the Micro-politics of Corporate Environmentalism. *Organization Studies*, 33: 1451–1475.

Chapter 21

# Turning Towards the Tears of the World: Practices and Processes of Grief and Never-endings

*Jo Hamilton*

## Climate Change and Grief

*You can only stand the silent scream*
*of the picture that is forming*
*for so long. You can almost not sleep.*
*Amazing things can happen when feeling the connection.*
*Opening perspectives, sharing grief, alchemising into passion and deep determination.*
*You can reimagine yourself and the earth.*
(selected phrases from participant interviews arranged by the author)

There are a range of emotions and affects connected to climate change, which are experienced by different publics at different times. They include grief, love, hope, hopelessness, anger, fear, guilt, alongside a numbness or absence of feeling. Whilst these emotions and affects are interconnected, in this chapter I focus on grief.

Grief can relate to the actual or anticipatory loss of people and things that are loved and valued, and the injustices associated with these losses. A willingness to face the reality of a loss is an important part of processing grief (e.g., Kübler-Ross' 'stages of grief', 1970), and the non-linear 'tasks of grief' (Worden, 1991). These tasks cover 'accepting the reality of the loss'; 'working through the painful emotions of grief', 'adjusting to the new environment/ developing a new sense of self' and 'reinvesting emotional energy' (Randall, 2009: 122–123). These processes can contribute to a healing process whereby painful feelings don't necessarily disappear, but the relationship to them changes. In contrast, not acknowledging or working through grief can contribute to melancholia or forms of depression (Lertzman, 2015).

Western grief theories (Kübler-Ross, 1970; Worden, 1991) centre around confronting one's own death, or the death of a loved one, which usually have an end point. More holistic approaches to grief, such as Weller's 'gates of grief' (2015), draw on Buddhist philosophy and Dagara wisdom traditions of Burkina Faso (Somé, 1995), and acknowledge the ongoing griefs to be lived

DOI: 10.4324/9781003436096-38

with. Weller's 'gates of grief' typology include losses of people and things that are loved, places that have not known love, the sorrows of the world (relating to socio-ecological issues), grief associated with what we expected but did not receive (e.g., parental love, or a connection to the more-than-human world) and ancestral grief (Weller, 2015).

In the global north, a range of socio-ecological[1] and personal griefs are connected to climate change for those who are engaged and concerned (Randall, 2009; Lertzman, 2015; Cunsolo and Ellis, 2018; Head, 2016). These include grief for past, present and anticipated future losses of people, places and ecosystems, losses of environmental knowledge and identity (Cunsolo and Ellis, 2018) and current and/or anticipated grief for losses of existential safety, and threats or implied losses to current lifestyles (Randall 2009; Head, 2016). Cultural contextualisation is needed, as for many people the past and present, let alone the future, is not characterised by safety or a comfortable lifestyle (Whyte, 2018).

Grief needs to be acknowledged and worked with to enable and sustain forms of engagement with climate change, as Akomolafe (2023) asks 'what if grief is the invitation and not the impediment?'. Processes to acknowledge grief can 'strengthen people's capacity to be in their own and with others' distress' (Moser, 2015), and can support active engagement to 'remake our futures using all of our creativity, reason, feeling and strength' (Randall, 2009: 128). Head argues that acknowledging the griefs associated with climate change needs to be an 'explicit part of our politics' (2016: 2), whilst Akomolafe (2023) suggests that 'grieving together, falling apart together might very well be the most ecstatic, the most animated politics in response to these moments that we can master'.

Whilst grieving and mourning are associated with anticipated or actual losses, melancholia can be defined as a state which arises when 'one cannot see clearly what it is that has been lost' (Freud, 1917: 245). Melancholia can present as a loss of interest in life, a stuckness, a loss of capacity to love and a withdrawal. Melancholia and depression have been felt in connection to the politics of climate change, through feeling a loss of hope or possibility of change (Osborne, 2018), or an ambivalence arising from conflicts between attachments to a lifestyle: of both being part of a system that causes harm, and wanting to mitigate the damage (Lertzman, 2015). This can contribute to depression and a withdrawal from individual or collective forms of engagement with climate change. Whilst such a withdrawal can be an adaptive coping strategy in the short term, in the longer term it could be seen as a maladaptive coping strategy as it prevents the exploration of the root causes (Andrews, 2017).

## Focus of Research

This chapter draws on doctoral research[2] conducted in England and Scotland between 2017 and 2019. Interviews were conducted with 17 participants in two group work approaches that incorporated acknowledgement and/or expression of emotions connected to socio-ecological issues: the Work that Reconnects/ Active Hope and the Carbon Literacy Project (CLP). The Work That Reconnects (TWTR) is an evolving body of experiential group work practices developed in the USA, Europe and Australia onwards by Joanna Macy and colleagues (Macy and Brown, 2015). It draws on Buddhist philosophy, systems theory and deep ecology. Active Hope (AH) (Macy and Johnstone, 2022) draws on the TWTR but is offered in book and online forms. Here I refer to TWTR/ AH together. Both primarily – but not exclusively – attract those with an existing degree of interest in and engagement with socio-ecological issues, and both have been integrated into many climate change engagement approaches.

The Carbon Literacy Project (2020) offers training in carbon literacy, typically in workplace contexts. It assumes no prior knowledge of or active engagement with climate change and is delivered via a day-long workshop by trained facilitators. At the time of research, the workshops provided space to acknowledge and reflect on participants' emotional responses to climate change but did not focus on exploring emotions in more depth.

Both TWTR/AH and CLP were traditionally offered in person. With the expansion of online options and COVID-19 restrictions limiting in-person interactions, TWTR/AH has increasingly been offered online as a one-off or as a series of workshops with accompanying exercises, and CLP now offers e-learning courses.

Participant interviews were conducted and analysed, drawing on psycho-social methods, including Biographical Narrative Interpretive Method (BNIM, Wengraf, 2017) and free association narrative interview (Hollway and Jefferson, 2000). The participant interviewees were aged between 24–75, all were racialised white and had pre-existing concerns about climate change. Their engagement ranged from being concerned and taking forms of private-sphere actions (e.g., domestic energy and carbon reduction) to being involved in public-sphere actions, such as civil-society organisations (CSOs) for over ten years. Participant names in this chapter are pseudonyms.

## Grief Connected to Climate Change

Three main themes of grief emerged through the participant interviews. These connected to acknowledging, experiencing or witnessing the injustices of loss of ecosystems, peoples and cultures; loss of lifestyle, expectations and possibilities; and loss of connection to the more-than-human world.

Table 21.1 gives example quotations to illustrate participants' connections between grief and climate change, both generally and experienced in TWTR/ AH and CLP workshops.

The *loss of peoples and ecosystems* theme was connected to the impacts and associated injustices of social and ecological destruction, both close to home *and* spatially distant. For some, witnessing impacts near home linked to the magnitude of global impacts.

*Table 21.1* Examples of grief themes

| *Theme* | *Participant* |
|---|---|
| Loss of peoples and ecosystems | *It was really helpful and really powerful for me to be able to go up and have space to express that … I just broke down in tears … talking about ecocide and the loss of biodiversity* [Nic, TWTR/AH]. |
| | *The die-off [of nesting sea-bird chicks] was horrendous, it was like watching these mini-massacres happening constantly* [Katrina, TWTR/AH]. |
| Loss of lifestyle, expectations and possibilities | *If someone were to say … you can't go to the Far East ever again … I would grieve for that, because I like going to new places and doing new things* [Sally, CLP]. |
| | *All that stuff makes you sad, because you have a … childhood fantasy of what your life will look like, and then actually it's quite different* [Sally, CLP]. |
| | *I also want to have something for myself … how far can we go [reducing carbon emissions] without making each other miserable?* [Tom, CLP]. |
| | *To think that I can't change the world … sharing the grief of not having the answers, and the frustration and powerlessness it brings, is … empowering* [Bela, TWTR/AH]. |
| Loss of connection to more-than-human world | *In our culture we've become a culture of empire … it's kind of like a dominant narrative … we've done so much dominating in the past it's as if nature is something we dominate … there's so much destruction to that in every way* [Kate, TWTR/AH]. |
| | *I had just lost contact with my childhood … So many people who have lost experience and … the ability to be with other species, and that's what lies behind the extinction crisis. … that's part of my lost childhood, and it's finding the strength to be a child again* [Ben, TWTR/AH]. |

The *loss of lifestyle, expectations and possibility* theme presented in different ways. For those taking private-sphere actions, anticipatory grief was apparent when considering giving up – or losing a guilt-free enjoyment of – aspects of their lifestyles they cared about and valued. This was linked to the assumption of low-carbon lifestyles as being 'miserable', and was mostly expressed in muted, resigned or defensive ways. The loss of expectations was also articulated by two participants questioning whether to become a parent, indicating a greater depth of potential loss. The loss of possibility connected to not having the answers and feeling frustrated by the lack of political action on the scale needed.

The *loss of connection to the more-than-human world* theme was expressed by five participants, illustrating how climate change materialised the ongoing grief of separation and loss of connection. This was experienced individually and culturally. For example, Ben's loss of connection to nature was shaped by being cut off from his love of nature from childhood, whilst Kate's grief reflected the injustice of empire-dominating cultures and the natural world.

Enfolded in all three themes were griefs from the participant's life stories. In most cases participants felt able to explore these in TWTR/AH spaces. However, one TWTR participant reflected that because of the level of deep personal grief they were holding at a workshop, the workshops felt too overwhelming for them at the time.

There was a relationship between how grief arose in participant interviews and their degree of climate change engagement. For those engaged in private-sphere actions, grief was associated with known and anticipated losses, alongside the implications of that knowledge on their lifestyle and expectations. The latter was set in binary opposition to the negative associations of green lifestyles and contributed to forms of ambivalence. For those involved in public-sphere actions, grief was expressed more viscerally, connected to the knowledge of socio-ecological impacts, and anger and frustration about inadequate political and systemic responses.

## How Grief Was Worked within TWTR/AH and CLP

TWTR/AH participants acknowledged and experienced grief connected to climate change in workshops in a range of ways. These included a surfacing of feelings that may have been occluded or not expressed publicly, their grief being witnessed and by recognising and empathising with grief expressed by other participants. They reflected on their emotions such as grief, fear and anger through movement-like qualities, such as *turning towards*, *facing* and *embracing*.

For most TWTR/AH participants, the opportunity to work with grief contributed to them both becoming engaged – and sustaining an active engagement – with climate change. I clustered these into four themes, shown in Table 21.2 – healing relationships; deep determination and resilience; sharing the grief; and more-than-human connections. Some responses reflect participation in one workshop or program, whilst others reflect participation in a range of workshops over time, supported by practices such as mindfulness, movement and meditation.

The *healing relationships* theme illustrates how the space to explore emotions such as grief enabled a healing relationship to oneself, a reconnection within. The *deep determination and resilience* theme illustrates how participants developed inner resources in relation to issues they care about, in this instance climate change. S*haring the grief* illustrates the importance of feeling connected to a wider community, and witnessing shared emotions. This was particularly evident during more ceremonial or ritual exercises where emotions were visible, and experiences of inter-vulnerability became sources of strength and connection. The *more-than-human connections* theme illustrates how participants experienced regenerative relationships with the more-than-human world.

*Table 21.2* Relationship to engagement and agency

| | |
|---|---|
| Healing relationships | *I've got this resource, this tool … that gives me that feeling of safety* [Nic, TWTR/AH].<br>*enabled the healing of some of the pain* [Ben, TWTR/AH].<br>*Connection to myself that I felt that I'd lost over the years* [Lucy, TWTR/AH].<br>*an exploration of my inner world … led to this feeling of connection, in terms of global problems and issues, it sort of was the catalyst … my engagement in climate issues has been completely coloured by that aspect* [Evie, TWTR/AH]. |
| Deep determination and resilience | *I'd connected to a deep determination and passion to continue the work* [Nic, TWTR/AH].<br>*I haven't come across anything other than [Active Hope] that actually gives a positive way of navigating now through this* [Kate, TWTR/AH].<br>*I don't get frantic anymore. I'm much calmer in a way* [Helen, TWTR/AH].<br>*I've noticed an improvement in my resilience* [Lucy, TWTR/AH]. |
| Sharing the grief | *allowing myself to rest back in everybody's arms and share that grief is immensely powerful* [Bela, TWTR/AH].<br>*I remember one particular process … [Joanna Macy] put two huge bowls down in the conference hall … people coming down and sharing and giving … these things became the tears of the world … it was just a beautiful connecting process* [Robert, TWTR/AH].<br>*the other person looked at you and recognised your grief … collectively … connecting with each other about that … was really powerful* [Kate, TWTR/AH].<br>*I'm not alone in this time of converging crises … dissolved feelings of isolation* [Nic].<br>*it's not just my work that is going to make a difference* [Helen, TWTR/AH].<br>*feeling that connection with those people* [Evie, TWTR/AH]. |
| More-than-human connections | *if you let yourself be channel, a conduit, for rethinking, yourself, and having the courage to believe it, you can reimagine yourself, the earth, you knew it the way you were born* [Bela, TWTR/AH].<br>*see more of the feminine to help with coming back to respecting the land, and with that being acts to address the climate emergency [and] to love and respect the land, to come back to relationship in an honouring way* [Kate, TWTR/AH].<br>*generating … a different kind of relationship with other species, with the whole of life around, getting outside the Anthropocene* [Ben, TWTR/AH].<br>*I'm a zoologist, but it became theoretical rather than practical. And it was a good reminder of how it's there for a reason, I need that, it really is part of me rather than a title* [Lucy, TWTR/AH]. |

In the TWTR/AH workshops, working with grief required a safe-enough space, a strong-enough container, permission to share a range of griefs (alongside other emotions) and a validation that came from witnessing that others may feel similarly. Whilst the facilitator was holding the space, it was also held by the group and through practices of connection with the more-than-human world. When some of these elements were not present, some participant interviewees felt too overwhelmed to bring their grief. This highlights the importance of attention to trauma-informed group processes, ensuring that exercises are invitational, and respecting participant's boundaries and choices about the depth of emotional exploration undertaken, and with whom.

In contrast, the CLP workshops offered acknowledgement of emotions such as grief, but not a space to unpack and work through them. This is understandable given a workplace context focused on climate change: participants bring differing degrees of familiarity with their emotions and may not wish to explore emotions with colleagues. However, all four CLP participants displayed forms of ambivalence, which prevented further exploration of their grief. Their internal vacillation was revealed through themes of 'knowing and not knowing', caught between 'future hope and hopelessness', and considering forms of action to be 'all or nothing'. (Table 21.3)

*Table 21.3* Forms of ambivalence

| *Theme* | *Illustrative quotes* |
|---|---|
| Knowing and not knowing | *there were definitely ugly truths that we knew about but we kind of ignored. Stuff like we like travelling a lot* [Sally, CLP].<br>*I thought, I'm here to have a good time, this is a once in a lifetime thing …I never really followed it on if I'm honest. I think I felt that it was just so big* [Tom, CLP].<br>*you can almost not sleep about it because it gets to that point* [John, CLP]. |
| Future hope and hopelessness | *I think about it all the time … I go from feeling empowered … to just feeling pretty hopeless … sometimes I feel 'man, if we get this right, I mean if we can fix it, then we can fix anything'* [Sally, CLP].<br>*you don't want to think that one day your great [descendants], or great great [descendants] are going to be like 'oh that person, or they're responsible … You don't want that to be the legacy when you lived. So … if that did happen, God forbid that it doesn't, but if it does you want to at least be on the right side of it* [John, CLP]. |
| All or nothing | *I think we can all make a difference, and that doesn't necessarily mean becoming like an eco-warrior … if you do own a car, or … are taking a flight … then you're an enemy of the state* [Tom, CLP].<br>*all or nothing, so it's kind of 'right you have to kind of give up your car, and never fly again, and um become a vegan and make your own clothes, or you're just a bad person* [Sally, CLP]. |

The theme of *knowing and not knowing* reveals patterns of ignoring or holding knowledge at bay. It was connected to fears of what participants might lose, and their perception of lack of agency. The *two stories of the future* theme revealed an ambivalence between two oppositional stories of the future, that operated as a future-oriented pendulum. The theme of *all or nothing* demonstrates the polarities of being an *eco warrior* or *enemy of the state*, and provides insight into the associations, conflicts and tensions associated with a 'perfect standard' of environmentalism and climate change engagement.

An important aspect of both the TWTR/AH and the CLP is an openness to the outcomes. Whilst workshops were framed by climate and wider socio-ecological issues, they were not prescriptive about *how* people should engage. The CLP encouraged active engagement, but participants decided how they would take this forward. TWTR/AH helped develop agency and supported participants to explore and develop their relationship to active engagement. For some this provided resources for deeper engagement, whilst others acknowledged their overwhelm and burnout, enabling them to step back from active engagement to resource and take stock.

## Discussion

My findings illustrate how climate change materialised actual or anticipated losses relating to what participants value/d and care/d about in the past, present and future. Much of the results reinforce existing literatures (Randall, 2009; Head, 2016; Cunsolo and Ellis, 2018; Norgaard, 2011; Lertzman, 2015) by adding examples of how grief was experienced and worked with in two group work approaches in England and Scotland.

TWTR/AH provides examples of *how* the productive generative and reparative potential of grief could be realised, providing ingredients for Akomolafe's suggestion of an *animated politics*. The TWTR/AH provided contexts for the tasks of grief (Worden, 1991) to be undertaken and to explore the connections between different 'Gates of Grief' (Weller, 2015). As Worden emphasises (and Randall, 2009, reinforces), these 'tasks' are optional. They required participants to attend a workshop, to be prepared to encounter the associated emotions and affects, and to continue their exploration. Some participants used TWTR/AH to work with their grief in an ongoing way, developing their capacity to inhabit the 'long emergency' (Hasbach, 2015).

The CLP participant interviews illustrate the importance of incorporating acknowledgement of emotions in workplace climate change engagements. At the time of interviews, participants had not had the opportunity to explore their emotions in more depth, particularly those connected with loss of lifestyle, expectations and possibility. Thus, despite their expressed concern and desire to be more actively engaged, defences were apparent which impeded both their engagement and further exploration of the uncomfortable paradoxes of privilege, threats to identity or working through the emotions they associated with an unbearable future. This highlights the importance of acknowledging the range of emotions connected to anticipated losses of lifestyles and expectations and offering a range of spaces and resources after workplace workshops – or other forms of engagement. This could include follow-up workshops, peer-to-peer mentoring or signposting to opportunities to explore the issues in more depth.

Whilst I focus on grief in this chapter, my research illustrated the connections between emotions experienced by participants in relation to climate change: love was enfolded within grief; fear of loss derived from grief; and anger was one outward expression of grief. Aspects of grief were connected to degrees of active engagement with climate change and were expressed in relation to participants' biographies and lives. This highlights the importance of acknowledging the range and depth of griefs enfolded in climate change, and the interwoven connections between personal and socio-ecological griefs.

The TWTR/AH and CLP are two contrasting approaches to public engagement with climate change. Both offer insights and perspectives on how grief is experienced in relation to climate and wider socio-ecological issues, and in the case of TWTR/AH, how grief can be worked with in these contexts. The research has emphasised the importance of attending to and acknowledging the granularity of emotional responses to climate and socio-ecological issues, alongside the accompanying tensions and ambivalence, and to provide opportunities and resources for these to be explored in more depth.

## Notes

1 I use the term 'socio-ecological' as feelings about climate change exceed the losses from climate impacts and include the accompanying social injustices which disproportionately affect those who have been marginalised and oppressed associated with the causes (e.g., extraction of fossil fuels), impacts (e.g., floods and extreme weather) and mitigation approaches (e.g., impacts from the extraction of rare earth metals for renewable energy components).
2 Permissions given by participants in PhD research on which this chapter is based.

## References

Akomolafe, B. (2023) *This Is the Part Where We Fall Down: On Climate Grief and Hope.* Video of online event 4 May 2023 with Naomi Klein, Bayo Akomolafe, and Yuria Celidwen. Othering and Belonging Institute. Available at: https://news.berkeley.edu/2023/06/02/berkeley-talks-transcript-climate-grief-and-hope/ [Accessed 30 June 2023].

Andrews, N. (2017) Psychosocial Factors Influencing the Experience of Sustainability Professionals. *Sustainability Accounting, Management and Policy Journal*, 8(4): 445–469. 10.1108/SAMPJ-09-2015-0080.

Carbon Literacy Project, 2020. *The Carbon Literacy Project.* [online] Available at: http://www.carbonliteracy.com/ [Accessed 12 February 2020].

Cunsolo, A. and Ellis, N. R. (2018) Ecological Grief as a Mental Health Response to Climate Change-related Loss. *Nature Climate Change*, 8(4): 275–281.

Freud, S. (1917) Mourning and Melancholia. In: J. Strachey (translator) *On the History of the British Psycho-analytic Movement*, Volume XIV 1914–1916. London: Hogarth Press.

Hasbach, P. H. (2015) Therapy in the Face of Climate Change. *Ecopsychology*, 7(4): 205–210.

Head, L. (2016) *Hope and Grief in the Anthropocene: Reconceptualising Human Nature Relations*. London: Routledge.

Hollway, W. and Jefferson, T. (2000) *Doing Qualitative Research Differently: Free Association, Narrative and the Interview Method.* London: SAGE.

Kübler-Ross, E. (1970) *On Death and Dying: What the Dying Have to Teach Doctors, Nurses, Clergy & Their Own Families.* New York: Macmillan Publishing Company.

Lertzman, R. (2015) *Environmental Melancholia.* London: Routledge.

Macy, J. and Brown, M. (2015) *Coming Back to Life. The Updated Guide to The Work That Reconnects.* Gabriola Island, BC: New Society Publishers.

Macy, J. and Johnstone, C. (2022) *Active Hope (revised): How to Face the Mess We're in with Unexpected Resilience and Creative Power.* Novato, CA: New World Library.

Moser, S. (2015). Whither the Heart (-to-heart)? Prospects for a Humanistic Turn in Environmental Communication as the World Changes Darkly. In: A. Hansen and R. Cox (eds) *Handbook on Environment and Communication.* London: Routledge, Ch. 33.

Norgaard, K. M. (2011) *Living in Denial: Climate Change, Emotions and Everyday Life.* Cambridge, MA: MIT Press.

Osborne, N. (2018) For Still Possible Cities: A Politics of Failure for the Politically Depressed. *Australian Geographer*, 50(2): 145–154. doi: 10.1080/00049182.2018.1530717.

Randall, R. (2009) Loss and Climate Change: The Cost of Parallel Narratives. *Ecopsychology*, 1(3): 118–129. doi: 10.1089/eco.2009.0034.

Somé, P. D. (1994) *Of Water and the Spirit. Ritual, Magic and Initiation in the Life of an African Shaman.* New York: Arkana.

Weller, F. (2015) *The Wild Edge of Sorrow: Rituals of Renewal and the Sacred Work of Grief.* Berkeley, CA: North Atlantic Books.

Wengraf, T. (2017) Biographic-Narrative Interpretive Method (BNIM). Short Guide bound with the BNIM Detailed Manual. Interviewing and case-interpreting for life-histories, lived periods and situations, and ongoing personal experiencing using the BNIM. Current updated version available tom.wengraf@gmail.com

Whyte, K. P. (2018) Indigenous Science (Fiction) for the Anthropocene: Ancestral Dystopias and Fantasies of Climate Change Crises. *Environment & Planning E: Nature and Space*, 1(1–2): 224–242.

Worden, J. (1991) *Grief Counselling and Grief Therapy.* Second Edition. London: Tavistock.

Chapter 22

# The Psychological Work of Being with the Climate Crisis

*Chris Robertson*

Whether we talk of tipping points, ruptures or paradigm shifts, it is becoming increasingly clear that humans are at a geological and cultural threshold as the earth shifts from its stable Holocene period to an unstable one, sometimes referred to as the Anthropocene. Symptoms of this threshold appear in consulting rooms and therapeutic groups. This chapter focuses on the psychological work involved in addressing these collective symptoms, such as eco-anxiety, disorientation, terror, despair, helplessness and rage.

To address the cultural pathology of a society, seemingly bent on destroying the fundamentals of life support on our planet, requires stepping out of the context of individual pathology. This chapter attends in particular to the workshop *Through the Door* that the Climate Psychology Alliance (CPA) has been offering since 2018. It outlines the background of how the project has supported psychotherapists and other professionals, themselves struggling with incapacity and helplessness. In the second part, I share the work around therapeutic practices, which although familiar to psychotherapists, are especially fit for the challenges of this uncertain time.

## Background

In his article *From Mirror to Window (1989)*, American Jungian analyst and founder of Archetypal Psychology, Hillman urged analysts to stop gazing at the mirror's reflection and look through the window at the world outside. Climate activists would say that looking is not enough; there needs to be engagement. This tension between psychological reflection and political action has a long history that can emphasise the polarity. I will be exploring a third point between the psychological containment of overwhelming ecological anxiety and the desire to 'do something'. This psychological work is an engagement with unconscious complexes together with a reimagined vision of life on a damaged planet.

Going *through the door*, be it a real consulting room door or a symbolic portal, evokes a passage that requires relinquishing the 'normal' and opening to another reality. The transit marks a passage, such as at mid-life, in need of

DOI: 10.4324/9781003436096-39

radical reappraisal. To be able to pass through the door requires relinquishing attachment to the known order and being present to an emergent but as yet unknown reality. This process of relinquishment, of stripping away, is common in many traditional rites of passage, in which the old self is symbolically shed like an old skin.

To leave the consulting room does not deny the value of the closed door of psychotherapy in creating a holding and incubatory space where both healing and dreaming can take place. Such a space is needed in an agitated and manic culture such as ours. A problem with this enclosure is that it can turn environmental problems into interpretations of an internal world. Much anxiety necessarily accompanies the social fragmentation and cultural rupture of the climate crisis. This needs a place to be heard.

In her article on the collective response to the climate emergency, Steffi Bednarek (2019) writes

> whilst there is no doubt that some people will need the safety of one-to-one support and the clinical expertise of a well trained psychotherapist, others may need community as an antidote to the extreme individualism that we have all been subjected to. After all, a collective wound may require collective healing.

The *Through the Door* workshop is not a solutions-oriented skills training for practitioners but a learning to bear what feels unbearable and think the unthinkable. It is an experiential stepping across the threshold of the consulting room to explore what permits the emergence of fresh responses to the challenges of this destabilising time.

In an early workshop, a senior psychotherapist protested that her patients never broached climate issues. Rather than leaving her with defensive guilt, we used her protest as a starter for the group to explore how they gave permission for patients to bring their concerns and distress about climate and environmental destruction. This giving of permission for clients to speak about their unthinkable anxiety is not straightforward. It cannot be mandated. It needs a letting go of control and expectation while at the same time having therapeutic antennae listening out for a client's hesitant broaching of climate-related discomforts.

In this way, permission for difficult climate conversations resembles making space for other taboo subjects, such as sexual abuse, that may have been perceived as implicitly unwelcome within the psychotherapy. What can make it safe enough for these conversations to be risked? Within a group setting there is a gradual recognition of a shared ambivalence and a relief in taking the plunge to speak out any anxieties. The sharing of desolation brings a solace if not a solution or remedy. The group itself gives permission to bring these risky topics and slowly adapts to this new sense of vulnerability. The shared sense of support makes it possible to engage these dilemmas that confound and overwhelm us separately.

In individual psychotherapy, it is not the inability of the psychotherapist to bear the emotions, so much as the failure of the container to carry the waves of collective anxiety which inhibit engaging with climate-related emotions. A client may be ambivalent about bringing this dangerous and taboo-like material into the consulting room. It can feel too dangerous to risk. An unconscious collusive alliance can form to protect the therapeutic relationship from this collective danger. A shift away from attributing the problem to the individual and instead seeing the sickness in the community is a vital move. It is not the individual or their family but the life-world of the community that is dis-eased.

Climate conversations often carry a heavy charge that make them no-go areas in social situations. While this is beginning to change as the culture wakes up to the widespread climate disasters, it is worth considering what is at stake in speaking out or not. Something gets evoked that makes climate conversations awkward, as if an attack or persecutory accusation might follow. The fear can be that any acknowledgement that all is not well is likely to lead to rejection. The privilege of belonging is not having to own the discomfort. Attempting to speak out may feel like a betrayal of bonds that tie our social identity into a jointly constructed 'derangement' (Ghosh, 2016).

Opening to what is 'other' beyond the doorway carries dangers of disrupting the escapist consumerism that fills the nihilism of modernist life. It is like a threatening nightmare that has erupted into waking consciousness through talk of climate crises and the end of the world. These ruptures of the basic assumptions that underlie how the fossil-fuelled privilege has been permitted to construct a neo-liberal world are disruptive and scary, but potentially vitalising.

## Practices for an Emergent Reality

In an enquiry into the longer-term effects of participating in the *Through the Door* workshop, we are finding that liminal states evoked in the workshop are vital in evoking new ways of seeing different landscapes and different stories. Participants have found a fresh attunement that brings with it unexpected opportunities and synchronicities between internal states and outer reality.

I shall focus on these practices:

- becoming porous, permeable, and vulnerable
- negative capability – staying with the discomfort
- acceptance and remorse
- imaginal reality and storytelling

## Vulnerabilities

Participants in the workshops often feed back their surprise at its experiential nature. Going into how the climate crisis impinges on them is disturbing.

Witnessing others' deeply felt concerns amplifies this. These may be concerns about industrial farming, pollution that is killing river life, social injustice or a sense of marginalisation. Often these exchanges move from rage and complaint to acknowledgement of inadequacy, failure, guilt and shame. Rather than a cosy conversation, it is as if another reality has broken in, as participants exchange previously suppressed and deeply felt concerns, such as ecocide.

There is a parallel process that affects the workshop facilitators. The visceral power of the emotions upends clear plans or expectations. They may no longer feel in control without knowing what is meant to happen. This evokes vulnerability, echoing times of feeling lost or confused. A floundering of our personal ego touches personal wounds, and the climate context of the group evokes ecological wounds. Within individual work, this parallels the psychotherapists' urge to bracket their own subjectivity in order to maintain control. Much as the ego and its defences might want us to believe in our separate independence, the price for being invulnerable is alienation and isolation. The focus of the workshop bridges the shared nature of this wounding, reminding us of our connection and mutual dependence.

If the walls of the consulting room are held as semi-permeable membranes, they offer a porous containment to the troubles of the world. Being porous and permeable sensitises humans to the suffering of the other-than-human. Within a group workshop, where the shared vulnerabilities attune members to the suffering outside of the group, this permeability becomes potent. The crisis is not something happening out there. We are entangled with the natural world so that its destruction (at human hands) is our devastation too.

As one participant put it:

*Lots of images and thoughts drifting back and forth. It felt fertile and productive but also painful and confusing. But, above all I felt energised. I don't think it's made me feel clearer about what I should be doing, or how, but I feel inspired and alive to possibilities! The group thinking triggered not just the loss and chaos of the Crisis but also my personal story. I think this was what the 'powerful' connection enlivened in me.*

Recognising our mutual vulnerability is often what brings communities together. We know that we need each other. The mutuality evoked in the Through the Door workshop is extended to that between humans and the other-than-human. We are reminded of Abram's (1997) 'reciprocal affinity' between ourselves and the 'others' of the planet, including its biosphere, that comes pre-attuned through co-evolving together.

Part of the workshop involves leaving the room and going out into external nature. Inevitably participants meet other-than-humans, and these encounters often bring heightened feelings of loss that can be difficult to name. A sensitivity to ecological harm was given a name by Australian philosopher Glenn Albrecht (2019). He called it Solastalgia – a kind of homesickness caused by environmental change such as when a familiar place is rendered unrecognisable, for example, by natural disaster or human destruction; the home becomes unhomely. The creating of a new word, such as Solastalgia, is an example of the value of naming troubling feelings so that they can be better understood.

## Negative Capability

Negative capability (Keats, 1958 [1817]) is a practice that inhibits the dispersal of anxiety through rushing to action. This type of holding and refusal to act prematurely has been a significant learning in the evolution of the Through the Door workshop. At the start of offering the workshop, we (the leaders) were concerned with providing participants with tools to take back into their work. Given psychotherapists' own anxieties about managing their clients' ecological distress, we wanted to offer practices they could work with. What we learned was that, although suitable skills for down-regulating and soothing have value, the deeper challenge is to acknowledge what we don't know about managing ecological distress. In fact, the value of sharing failures in the consulting room generated more creative dialogue than any apparent solutions.

In a culture where performativity, goal setting and production are heavily emphasised, the capacity to be receptive and stay with the trouble by tolerating frustration and anxiety is a vital practice. Within the workshop, making the space to digest and incubate, rather than moving on with the agenda is counter-cultural. This capacity to *be with* anxiety, to stay in the place of uncertainty, is what permits the emergence of fresh unthought thoughts. It is the practice of keeping open relational flow and resisting the mind's desire to bring premature closure.

Bion encouraged 'negative capability', which Keats (ibid.) described as 'when a man is capable of being in uncertainties, mysteries and doubts, without any irritable reaching after facts and reason'. Bion (1978) pointed out the temptation of familiar thought in prematurely closing off from what we don't want to see or hear. The capacity to be with the uncertain experience is especially pertinent at thresholds between phases, whether individual development, group maturation or cultural transformation. Other non-threshold phases may prioritise positive capabilities, such as clear leadership in a group. Naturally, Through the Door workshop groups vary, and some participants have wanted practical tips for managing eco-anxiety.

Bion also warned against 'filling the empty space' (1991) with preconceived solutions that obstruct new thoughts being recognised. Refusing the preoccupations and expectations of the controlling mind that wants to get somewhere or make something happen allows us to remain present to disturbing feelings. I imagine this attitude as like the gesture of extended hands, an empty cup, open to what is emerging.

## Acceptance

Acceptance is a feeling-based process of coming to *be with* how things are. This is more than mentally recognising the reality and is not to be confused with resignation, a passive defeatism that might perceive the world as doomed by the climate crisis. The process requires the work of relinquishing wishes that things could be other than they are. Letting go of these wishes for betterment is hard. It takes repeated rounds of relinquishing. It is an acceptance that there is no escape from our existential dilemmas, no recourse to saviours. Such *being with* is greatly supported by an empathic listener or group who understand the trouble from their own experience and does not dismiss, belittle or attempt to make it better. One participant in the TTD workshop commented that there was:

> *an absence of pressure on the leaders (or the group members) to come up with solutions but to support people in staying with, at times overwhelming, feelings and to offer support in living with the awful uncertainty about the future.*

The challenging reality of the climate crisis may result in not being able to work through the painful feelings. Once recognised, the scale of impending social fragmentation and potential collapse can feel beyond comprehension. It overrides our emotional capacity. Rather than attempting to get rid of these charged emotions through numbing or discharging, being with this overwhelm and the shame or guilt that it evokes is part of acceptance. Acceptance is not a final state or goal to achieve but a gradual process of being with what is, including our frailties and limitations. In terms of the apparent human inability to avert the climate crisis, this is a form of confronting. The acceptance of not being able to change it may paradoxically be what transforms the situation.

Although I am attempting to explicate the subtleties and radical nature of acceptance here, we do not teach this as an idea or separate practice in the workshops. It is an underlay of the work, a holding attitude that often catalyses open and creative conversations. A participant shared that she would 'not keep angsting about what more should I be doing about climate change … but to sit back (that is so difficult!) and trust that something will arise'.

One of the key obstacles to accepting our human situation with the climate crisis is the depth of anxiety and grief. When anxiety leads to apocalyptic fears of annihilation and human extinction, it is overwhelming. Attempting to face this alone can lead to depression, despair and suicide. In contrast, sharing these catastrophic fantasies within the safety of the group leads to their being witnessed and accepted. Having such feelings 'normalised' within the Through the Door workshop, it becomes safe enough to share anxieties not commonly expressed in the world. This can create a benign cycle in which knowing that the sharing need not trigger panic, builds the confidence to have these conversations.

The grief and remorse participants encounter are not resolved in the workshop. There may be guilt about complicity in the loss of the familiar world, but also remorse about what we have not done as bystanders. This brings new acuity in relation to climate issues, conversations with friends and family, and how to listen through to these concerns implicit within clients' stories. This resembles what Winnicott (1984) termed a 'capacity for concern' when the child becomes aware of their potential to hurt the mother and seeks to mend the damage. Winnicott says: 'Concern refers to the fact that the individual cares, or minds, and both feels and accepts responsibility' (ibid:72). Extending this to climate aware practitioners, it marks the transformation of guilt about human destruction of the environment to a sense of responsibility and the possibility of constructive actions.

## Imaginal Realities and Storytelling

A much-quoted statement from Thomas Berry (1978) emphasises that 'It's all a question of story. We are in trouble just now because we do not have a good story. We are in-between stories. The old story, the account of how we fit into the world, is no longer effective. Yet we have not learned the new story.' This is usually interpreted as an invocation for new stories. I prefer to attend instead to the notion that we are *in-between* stories. This in-between *is* where we are. It is uncomfortable being at the borders because the boundaries are confusing and unclear. The borderlands can be chaotic and unruly; they are also the spaces that are open to inspiration. Practical examples of being in-between are that twilight space between sleep and awakening, the phases of adolescence and mid-life.

This chapter started with tipping points and cultural thresholds where the instability of geological time and that of human society coincide. The threshold is balanced between the open and the closed, the inner and the outer. The Through the Door workshop builds on participants' own experience of resonance between inner and outer realities, leading to the notion that climate change is not just about 'solving a problem' but creating a new world. This 'new' is necessarily radically different from the old patterns of colonial and industrial exploitation. The workshop draws on imaginal practices to find new landscapes and languages to describe stepping through into this other world. A participant writes:

> *What was unexpected for me was a powerfully moving numinous experience in the guided meditation. So, my expectations of deciding on my next step really went out of the window, as I just needed to sit with what I'd experienced (and still am!).*

Henry Corbin (1969), French philosopher and Islamic scholar, used the term 'imaginal' to distinguish a meditative perception through receptive attention and reverie, from the *imaginary* as something made up. The *imaginal realm* has a presence of its own that offers images emerging through the collective unconscious that can carry intense meaning. These imaginal practices explore the pivotal edge between *letting go* and *letting come*; the letting go of the familiar known and the letting come of that emergent unknown. These both need courage. Courage to release control and courage to receive the power of a mysterious other. In the imaginal work, it is as if something or someone is waiting for us in this new landscape on the other side of the door. Some participants report that once they are through, what was invisible or hidden becomes natural, almost obvious.

Like dreams, such newly born experiences need to be held gently and nurtured. We have found that storytelling and free writing are confluent ways to amplify the precious experiences without concretisation or making them new ego goals. The free writing is without conscious control, with participants putting pen to paper and continuing without stopping whatever comes. The inspiration finds its way without 'help'.

In a way, much of the workshop involves participants telling stories to each other. The shift is from telling the story to having the story tell itself through the group's participation. Often the stories involve a wide social and cultural context in which new thoughts and proto-actions find a voice. There have been comments such as:

*We can't go on until everyone is with us.*
*Constellating a bigger story of what is within and between us.*
*Weaving a story together with all its different colours, shapes and mistakes.*

## End Echoes

The comments above highlight the group process of making a journey together – a collaborative journey of being-with the challenges and discomforts they brought with them. They echo the themes of this chapter. Facing into catastrophic ideas of a Western world threatening to unravel brings hard-to-bear feelings of loss and sorrow. Deep acceptance and remorse enable participants to unleash a capacity for concern, and allow previously unfelt reality to be imagined. The being-with engages participants in a reciprocal attunement of inner and outer. The workshop group mirrors the external cultural crisis and participants leave having a felt sense of the psycho-cultural thresholds.

The thresholds evoked through imagination in the workshop are both personal and collective. The cultural unconscious is constellated through the opening of a liminal space that is in-between the known and unknown. The cultural unconscious is less the denied cultural complexes stemming from collective trauma, and more the previously unimagined, unthought thoughts awaiting thinkers to apprehend them. The group's attunement sensitises participants to these future possibilities waiting to be perceived. Opening the door reveals a portal for renewal and rapprochement of inner and outer nature that has been split, through treating the earth as a dead and cheap resource to be exploited. The psychological work summarised here is a story of healing this alienating split.

## References

Abram, D. (1997) *The Spell of the Sensuous*. New York: Vintage Books.

Albrecht, G. A. (2019) *Earth Emotions: New Words for a New World*. New York: Cornell University Press.

Bednarek, S. (2019) 'This Is an Emergency' – Proposals for a Collective Response to the Climate Crisis. *British Gestalt Journal*, 128(2).

Berry, T. (1978) The New Story. Available at: http://newstories.org/wp-content/uploads/sites/8/2016/04/Thomas_Berry-The_New_Story.pdf [Accessed 29 April 2023].

Bion, W. R. (1978) *Four Discussions with W. R. Bion*. Strath Tay, Perthshire: Clunie Press.

Bion, W. R. (1991) *A Memoir of the Future*. London: Karnac Books.

Corbin, H. (1969) *Creative Imagination in the Sufism of Ibn Arabi*. Princeton: Princeton University Press.

Ghosh, A. (2016) *The Great Derangement: Climate Change and the Unthinkable*. Chicago: University of Chicago Press.

Hillman, J. (1989) From Mirror to Window: Curing Psychoanalysis of its Narcissism. *Spring*, 49: 62–75.

Keats, J. (1958 [1817]) in H. E. Rollins (ed.) *Letters of John Keats, vol. 1*. New York: Harvard University Press.

Winnicott, D. W. (1984) The Development of the Capacity for Concern. In: *The Maturational Processes and the Facilitating Environment*. London: Routledge.

Voice 12

# How does climate breakdown make me feel?

*Maddie Budd*

The deep sense of injustice. A darkness, a dread. Lost. Mad and alone. Disempowered.

I grew up in a rose-tinted world, a place of privilege carved out in bloody oppression and hierarchy. I didn't see this coming. It is a totally different reality. It is the end of the world.

This year I have done some extreme things. What else to do? I did them with some beliefs I'd rationalised of how they might shock people, how they might help people to see what's going on here. How just from one headline they might get people to feel the emotions that they pretend aren't there, break through that wall of emotional repression, denial. Maybe be the thing that tipped people to say, 'oh shit we've got to come together now'. But four months on, eight months on, I realise it was just as much about me, just as personal. *My* world is on fire. Me. My life. The older generations are pouring shit on *me*, on my life, on children. Or at least they are standing for that.

I thought it out: how can we get change here? What the hell kind of gamble might make people feel how serious this is, let this in? I felt alone enough to go it alone.

Even now my family, an immediate barometer, they act like none of it happened. They're not letting it in. They've done hardly a thing to act like this reality is real. To act like this is the end of the world. To fight. To fight for their kids. One time my dad said, 'you're not a child anymore'.

I feel let down. It takes bravery in me to say that.

For a lot of the past year, I was in fight or flight, and then I landed in prison for a few weeks, and it was like getting picked up by the armpits and still trying to throw punches. And then in the cell, all that grief that I'd not been letting myself feel, that I'd been forcing into the acts of sheer desperation that had become my everyday life, just flooded me. What else to feel? I lay on the wooden bench thinking if I'm in here for ten years it's not impossible that the officers will stop getting paid and we won't get fed, won't get let out. Or if I'm in here for two … then I've done my final action on this side of runaway breakdown. For that is how the maths adds up – a few years to avert runaway according to Sir David King.

DOI: 10.4324/9781003436096-40

In nonviolence, the reclamation of democracy, the only weapon we have is numbers. But with people continuing to deny this, with it being fully established that we are now beyond 1.5 into probable runaway warming, it felt like me against the world. The whole world on these shoulders. People would say come away from under the burden, you can't carry all that. And I'd say, to myself, in my heart, if I did that, all those children would be crushed. Please come beside me and shoulder it with me.

My dream is for actual solidarity, a realisation of our human potential for love, compassion, unity etc.

My fear is fascism, violence, and deep violation of human values.

And I suppose my real terror is the overwhelm of trying to live a life in collapse.

What I feel a lot at the moment, is a deep darkness. A deep dread, somewhere deep in my stomach. This deep ache in my organs. Flight, fight, freeze.

I've been feeling lost too: that's growing up, in a time where home is in exponential breakdown.

I feel walled in by paradox. Healing, wisdom, relationships – everything takes time – but we're in negative time.

I've felt mad and alone. I've clung to other people who are trying to disrupt the death-slide, who seem to feel the same. Then I didn't feel so mad or alone. That gave me the courage to do things that I felt I needed to. But there's a madness in that too.

I think it's all beyond believing in a way.

People talk about grief a lot. Of course, there's the grief. Grief for all the suffering, and all the suffering that is to come. All the beauty and dreams that are being stolen as extinction impends.

But what I have found so hard, is that this is a grief we are meant to feel together, with our communities, with our families. And yet that is not what's happening. Most people in my country, the UK, are not letting it in. My family, they are not letting it in. I think I know why: it is a huge psychological weight to bear. Perhaps a floodgate for other things they don't want to feel.

But in not feeling it now, we are shitting on this most brief, precious window of chance before climate breakdown becomes runaway. And in that there is responsibility. A responsibility to civilisation and humanity.

This is what people talk about as 'moral injury'. Seeing others fail in their highest moral duties.

So, in this way, it is a double grief. I want my family beside me, but they will not meet me where I am. I hope and long for unity, solidarity. For grieving with those I love, what we truthfully do need to grieve, and, together, to act.

A lot of young people are feeling this distress. Coming into life or coming of age into the biggest existential disaster, squared by the denial, the gaslighting, that is so pervasive. This difference which cannot be reconciled within our young minds. And nor should it have to be. We deserve better than this betrayal and lying. Then this mental health crisis which is the first wave of climate breakdown here in the global north – trying to live in a world that does not make sense. We deserve better than this injustice. We deserve better than being born into a humanity which stands by as it destroys itself, as young and vulnerable people die. Which is murder.

I wish millions of children weren't dying right now all around the world as growing conditions and safe water supply are decimated, as predicted. As ignored.

I am sad that in all this distress it can be so hard to remember my love for life.

It makes me feel fucking crazy. We are now living in a context of trauma. Of breakdown. It's hard not to feel death around every door. Tumbling, lost.

Alone even though we're all in this. But also, not always alone, connections are fierce nowadays. In fact, it's a mess of trauma from the actions I've taken, the places that's taken me, within the ongoing and worsening breakdown. Often, I feel panic and overwhelmed. I've been in deep despair; I have forced through feelings to extreme action and lived in constant flight or fight from being isolated and at odds with the system.

Sometimes people talk of a middle ground between satisfied despair and desperate hope: we have work to do.

People need to understand that this climate breakdown is the final waves of coloniality – the brutality of racism unravelling into a sickness of planetary scale. This brutality is nothing new, capitalism was honed through the enslavement of black Africans, the ideology of racism was fabricated to protect it, and the extraction and violence has only carried on since then. People have been finding ways to *live* within the death; people have been surviving colonialism for centuries.

To form a strong movement, we need each other: to grieve, for collective intelligence and wisdom, for power in numbers, for resilience, to not feel as crazy, for friendship, for the wisdom of eldership and the needs of youth. And to work to heal all these deeper wounds borne of a system built on violence and separation.

I am grieving my beautiful friend Xavi. My age, 22, and he was deeply in climate grief, I think he couldn't face being sent back to prison again.

All the beautiful kids. What the fuck. They're all going to grow to age totally left behind.

Millions are already dying.

My little sister and I were talking, fascism, history and history repeating itself, and climate breakdown. I said, 'in 2030 it's likely that 700 million people will be displaced in Africa because of drought'. She said, 'I'll be 21 then'. And something in her changed, like a shadow coming over her. And that put a fresh fear in me. I don't want her to feel this like I feel it. I don't want it to be ours to feel, or anyone's. Please look this in the eye. It's collapse of human civilisations, starvation, disease, war, fascism. The injustice and harm that's been going on for decades in places without law and order, in places being swept away, burnt to the ground.

Where's the humanity?

Where's the integrity?

'In the name of civilisation, four nations condemn Nazi forces of evil' – from the judgement at Nuremberg.

'No greater crime' – said by lawyers about the present-day climate and ecological death project. Out of all the evils, there is 'no greater crime'. And therefore, no greater moment of moral responsibility.

Maddie Budd, UK

# Name Index

# Keyword Index